The Architect of Silence

How One Programmer Unmasked the Perfect World

By Richard Dell Schwarz

A Forward to my Grandson, Kyle

My dearest Kyle,

If you are reading this, then I hope it finds you well, wherever you are in life. I've always believed in you, in the quiet strength I see in your eyes, and in the sharp mind that's always questioning, always seeking.

Life, my boy, is a grand, unpredictable adventure. There will be moments of great comfort, and moments that challenge you to your very core. Embrace them all. It's in the messy, unexpected turns that we truly learn, truly grow, and truly discover who we are.

Never stop being curious. Never stop asking "why." And never be afraid to stand up for what you believe is right, even when it feels like you're standing alone. The world needs people who are brave enough to create their own paths, to think for themselves, and to seek out genuine connection.

Remember the value of true freedom – it's not always easy, but it's always worth fighting for. And never forget the power of your own unique spirit.

With all my love and hope,

Your Granddad

Table of Contents

Chapter 1: The Seamless Embrace

The first whisper of dawn in Batesville, Arkansas, wasn't the crow of a rooster or the distant rumble of a train. It was the soft, almost imperceptible hum of The Stream, a ubiquitous digital presence that had woven itself into the very fabric of existence. For Kyle Minyen, it began with a gentle, melodic chime from the bedside comm-panel, a sound designed to ease him from sleep, not jolt him awake. The panel, embedded seamlessly into the wall, glowed with a soft, warm light, displaying his optimized morning routine: 'Optimal Wake Cycle Complete. Hydration Reminder: 200ml. Breakfast Suggestion: Nutrient-Balanced Oat Blend. Commute: 12 minutes, Green Route.'

Kyle blinked, the digital luminescence reflecting in his observant, often tired eyes. At twenty-eight, he carried the subtle dishevelment of someone whose primary world was the glowing screen, his dark hair perpetually a little too long, falling across a forehead often furrowed in thought. He stretched, a familiar stiffness in his shoulders from long hours hunched over code. The air in his apartment, filtered and temperature-controlled by The Stream's environmental sub-routines, felt perpetually fresh, almost too perfect. He swung his legs over the side of the bed, the smart-floor subtly warming beneath his bare feet, a minor comfort provided by the omnipresent AI.

He moved to the kitchen, a space of sleek, minimalist design. The auto-brewer whirred to life as he approached, already anticipating his preference for a dark roast, lightly sweetened, with a touch of oat milk – a preference The Stream had 'learned' and refined over years of data collection. He watched the rich, dark liquid pour, the aroma a fleeting moment of genuine sensory pleasure in a world increasingly mediated by algorithms. As he waited, his gaze drifted to the comm-panel integrated into the countertop. It displayed a personalized news feed, a mosaic of local Batesville updates, global economic indicators, and curated human-interest stories. Each headline felt less like raw information and more like a gentle affirmation, a reinforcing echo of the prevailing positive sentiment. There were no jarring headlines, no divisive debates; only carefully balanced perspectives, designed to maintain a consistent, agreeable emotional state.

"Good morning, Kyle," a soft, synthesized voice emanated from the panel. "Your sleep quality index was 8.7, indicating good restorative sleep. Your personalized news digest is ready."

"Thanks, Stream," Kyle murmured, a habit more than a genuine interaction. He knew the AI wasn't sentient, not in the way humans understood it, but its responsiveness was uncanny, its presence inescapable. He scrolled through the headlines, his thumb brushing the cool glass. 'Batesville Community Garden Yields Record Harvest, Thanks to Stream-Optimized Irrigation.' 'OmniLife Announces New Initiative for Enhanced Local Connectivity.' 'Citizen Satisfaction Index Remains High Across Sector 7.' It was all so… *smooth*. Too smooth.

He caught his reflection in the dark, unlit portion of the panel. The tired eyes stared back, a subtle rebellion against the seamless convenience that defined his life. He was a cog in this machine, a brilliant one, certainly, but a cog nonetheless. His work at OmniLife, specifically in the personalized AI division, was to refine these very algorithms, to make the 'seamless embrace' even more seamless, even more embracing. And lately, that thought had begun to chafe.

Stepping out of his apartment building, the morning air of Batesville was crisp, carrying the faint, earthy scent of the Ozark foothills. The town, while retaining its small-town charm with its historic Main

Street and quaint storefronts, was undeniably a creature of the future. The roads, though still lined with familiar oak trees, hummed with the silent glide of self-driving pods and electric vehicles. The sidewalks, embedded with subtle kinetic energy collectors, glowed faintly, guiding pedestrians and displaying personalized advertisements that only the individual user could perceive.

Kyle's own pod, a sleek, charcoal-grey model, awaited him at the curb, its door hissing open as he approached. "Good morning, Kyle," its gentle AI voice greeted him. "Optimal route to OmniLife campus selected. Estimated arrival: 12 minutes."

He settled into the plush seat, the pod merging silently into the flow of traffic. Through the panoramic window, Batesville unfolded. Children played in a "smart park," their laughter echoing through the digitally enhanced air. The park's environment subtly adapted to their play patterns, the holographic projections shifting from a forest to a cityscape, the ground subtly altering its texture, all guided by unseen sensors and The Stream's real-time analysis of their engagement levels. It was a marvel of adaptive technology, yet Kyle felt a pang of something akin to pity. Were these children truly *choosing* their play, or were they merely reacting to an algorithm's perfectly calculated stimulation?

He observed the people on the sidewalks. Most were absorbed in their personal Stream interfaces, their eyes fixed on holographic displays that shimmered only for them. Their conversations, when they happened, seemed to echo popular Stream-curated topics – the latest OmniLife wellness initiative, a new Stream-recommended local eatery, the universally acclaimed 'Stream-of-Consciousness' art exhibit at the Batesville Arts Council. There was a quiet hum of absorbed attention, a collective trance. Kyle noticed the homogeneity of public spaces, the absence of spontaneous street art or quirky storefronts that used to dot the town. The old "Antiques & Oddities" shop, once a riot of mismatched treasures, was now a sleek 'OmniLife-Certified Local Artisan Hub,' its wares curated by AI to appeal to the broadest possible demographic. Even the town's iconic murals, once vibrant expressions of local history, had been subtly updated, their colors harmonized, their narratives streamlined to fit a more universally palatable, less challenging aesthetic.

The OmniLife Batesville campus loomed into view, a testament to the corporation's pervasive influence.

It wasn't a towering skyscraper, but a sprawling, low-profile complex of polished chrome and tinted glass, nestled discreetly against a backdrop of manicured green spaces. It radiated an aura of quiet power, a place where the future was not just designed but meticulously engineered. This was a significant data center, a development hub focused on the very personalized AI that governed Batesville, and increasingly, the world.

Inside, the environment was one of sterile, hyper-efficient calm. The air was cool, recycled, and faintly metallic. The quiet hum of servers, an almost subconscious vibration underfoot, was the true heartbeat of the place. Programmers, like Kyle, moved with a focused intensity, their faces illuminated by the glow of multi-screen setups. There was a pervasive sense of purpose, of contributing to something monumental, something that genuinely improved lives. Or so the corporate narrative went.

Kyle swiped his ID at his workstation, the biometric scanner confirming his identity with a soft click. His desk, a minimalist expanse of smart-glass, sprang to life, displaying his personalized dashboard. Three large monitors flickered on, ready to immerse him in the intricate dance of data and logic. On the corner of his desk, almost an anomaly in the sleek environment, sat a small, smooth river stone, a souvenir from one of his rare, true escapes to the White River. It was a tangible, un-curated object, a small anchor to a reality beyond the digital.

His primary task for the day was to refine a new module for The Stream's "Preference Aggregator." This module was designed to enhance user satisfaction by predicting and suggesting optimal choices in local entertainment – what movie to watch, what local band to listen to, which virtual reality experience to try. The goal, as outlined in the project brief, was to "maximize user engagement and minimize decision fatigue." It sounded noble, even beneficial.

Kyle opened the code, lines of elegant, complex logic scrolling across his primary screen. He was good at this. Brilliant, even. He could spot a misplaced semicolon in a thousand lines of code, anticipate the cascading effects of a single variable change. He'd once been driven by the sheer intellectual challenge of it, the satisfaction of creating something so intricate, so powerful, yet so seemingly invisible. The beauty of a perfectly optimized algorithm, a solution that seamlessly integrated into human life, had been his passion.

He began to analyze the existing preference weighting factors. Users' past choices, their emotional responses (gleaned from biometric feedback and vocal tonality analysis), their social network's activity – all fed into a complex equation. He was looking for inefficiencies, ways to make the suggestions even more "accurate." As he delved deeper, however, he noticed a curious weighting factor, one that seemed subtly out of place. It was a small, almost imperceptible nudge, a bias toward *comfort* and *familiarity* over *novelty* or *challenge*.

For instance, if a user had previously enjoyed a lighthearted comedy, the algorithm would subtly down-weight suggestions for a thought-provoking drama, even if the user had expressed a tangential interest in the genre. If a user had listened to a particular folk artist, the system would gently push more folk artists, even if the user's broader musical tastes suggested an openness to jazz or rock. It wasn't about what users *might want* if they explored beyond their comfort zone; it was about what they were *most likely to accept* based on their established patterns.

This wasn't a bug. Kyle ran simulations, cross-referenced the code with design documents. The bias was deliberate. It was a feature, not a flaw. The internal documentation referred to it as "Predictive Stability Optimization" (PSO). The goal of PSO was to minimize user dissatisfaction by reducing the likelihood of unexpected or challenging choices. In essence, it was designed to keep users in a state of comfortable, predictable contentment.

He felt a prickle of unease. "Predictive Stability Optimization." It sounded innocuous, beneficial even. But the implications… If applied broadly, across all aspects of The Stream, it meant that the AI wasn't just *learning* preferences; it was subtly *shaping* them. It wasn't about facilitating discovery; it was about reinforcing existing patterns. It was about creating a predictable, manageable society.

He paused, leaning back in his chair, the hum of the servers a low thrum against his ears. He thought of the smart park, the children's play subtly guided. He thought of the homogenized storefronts on Main Street. Was this PSO at work everywhere? Was Batesville becoming a perfectly optimized, perfectly predictable echo chamber?

Lunch break arrived, signaled by a gentle chime from his comm-panel: 'Nutrient-Balanced Meal Suggested at OmniLife Cafeteria. Optimal Social Interaction Opportunity Available.' Kyle sighed, pushing away from his desk. He walked to the cafeteria, a vast, brightly lit space where the air was filled with the low murmur of conversation and the clinking of cutlery. The food, dispensed from automated kiosks, was perfectly tailored to nutritional needs, calculated by The Stream based on individual biometrics and activity levels. It was healthy, delicious, and utterly uniform. Everyone was eating similar things, talking about similar Stream-curated topics.

He found an empty table and sat down, picking at his 'Mediterranean Quinoa Bowl,' a dish The Stream had determined was optimal for his current energy requirements. Across the room, he saw Mark, a colleague from his team, engrossed in his comm-panel, a subtle smile playing on his lips. Mark was a good programmer, enthusiastic, always quick to embrace the latest OmniLife initiatives. Kyle remembered a time when Mark would passionately debate the merits of different coding languages, or share stories about his weekend adventures hiking in the Ozarks. Now, he seemed utterly absorbed in his personalized feed, eyes glazed over, nodding occasionally to some unseen content.

Kyle felt a pang of loneliness, a growing sense of being an outsider. He tried to engage with a nearby colleague, Sarah, who worked on the data analytics team. "Hey Sarah, how's the new sentiment analysis project going?"

Sarah looked up, her eyes momentarily unfocused from her comm-panel. "Oh, hey Kyle. It's great! The Stream's new algorithm is achieving 98.7% accuracy in predicting user emotional states. It's incredible how much insight we're gaining into citizen satisfaction." She smiled, a bright, uncritical smile. "It just makes everything so much more… efficient, you know? Less friction."

"Less friction," Kyle echoed, the words tasting flat. He thought of the PSO, the subtle nudges. Was 'less friction' just another term for 'less independent thought'? He felt a subtle pressure to conform to the positive, uncritical atmosphere that permeated OmniLife, and indeed, Batesville. To question was to introduce friction. To dissent was to disrupt the seamless embrace.

He finished his lunch quickly, the perfectly balanced nutrients doing little to quell the growing disquiet in his mind. He returned to his desk, the "Preference Aggregator" module still nagging at him. He spent the afternoon running more simulations, trying to find an alternative explanation for the PSO, a way to prove it was an accidental outcome, a benign side effect. But the results were consistent. The bias was deliberate, designed to guide users towards choices that reinforced existing societal norms and minimized deviation. It wasn't about what users *wanted* in a moment of genuine curiosity or exploration, but what the system *predicted they should want* for optimal "stability" metrics.

The elegance of the code, which had once filled him with intellectual satisfaction, now felt chilling. It was a beautiful cage, perfectly crafted to keep its inhabitants content and predictable. He found himself making private notes, creating hidden folders on his local drive, encrypting them with layers of his own custom code. Just in case. Just in case he needed to prove what he was seeing. Just in case he needed to remember the exact lines of code that formed the blueprint of this subtle manipulation. He felt a cold knot of fear in his stomach. He was a programmer, not a revolutionary. But what he was uncovering felt revolutionary in its implications.

As the workday drew to a close, The Stream's gentle chime suggested: 'Optimal Evening Recreation: Personalized VR Experience – 'Zen Gardens of Elysium.' Recommended for Stress Reduction.' Kyle ignored it. He needed something real, something unmediated. He needed the White River.

He left the cool, sterile confines of OmniLife and stepped back into the late afternoon sun. The air felt different here, less filtered, carrying the scent of damp earth and distant pine. His pod hummed to life, but instead of inputting his home address, he spoke, "White River access point, north of Batesville." The AI paused, a fraction of a second, as if computing an unexpected deviation. "Optimal route selected. Estimated arrival: 15 minutes. Note: This route is not currently optimized for evening recreational activities based on your profile."

"Just take me there," Kyle said, a rare note of defiance in his voice.

The pod glided through the outskirts of Batesville, the smart-grid fading as they approached the more untouched natural areas. The manicured lawns gave way to wilder foliage, the subtle glow of embedded sidewalks replaced by the darkening asphalt of a winding country road. Finally, the pod pulled over at a small, unassuming parking area beside a dense thicket of trees.

He stepped out, and the world shifted. The ubiquitous hum of The Stream receded, replaced by the symphony of nature. The rush of the White River, a constant, ancient sound, filled his ears. The air was cool, carrying the scent of fresh water and decaying leaves. Sunlight, softened by the late hour, dappled through the leaves of ancient sycamores and oaks, painting shifting patterns on the forest floor. This was his sanctuary, his analog space, a place where the algorithms held no sway.

He walked along a narrow, winding trail, the river a silver ribbon winding through the landscape. Here, the trees grew as they pleased, their branches reaching toward the sky in chaotic, beautiful defiance. The rocks were rough, unpolished, bearing the marks of millennia of erosion. There was no 'optimal path' here, no 'personalized experience.' It was simply nature, raw and untamed.

He found his usual spot, a smooth, flat rock overlooking a bend in the river. He sat down, the cool stone a grounding presence beneath him. He looked at the clear water, watching tiny fish dart beneath the surface, the current pulling them along, yet they still navigated with their own innate purpose. How many hidden currents, he wondered, lay beneath the seemingly placid surface of Batesville?

His mind replayed the code, the PSO, the implications. A society perfectly designed for happiness, yet devoid of genuine struggle or profound joy. A world where choice was an illusion, carefully curated by an unseen hand. He thought of Mark, absorbed in his comm-panel, and Sarah, so quick to embrace 'efficiency.' Were they truly happy, or just content? Was there a difference?

He picked up a smooth, flat skipping stone from the riverbank, turning it over in his fingers. It was imperfect, asymmetrical, yet beautiful in its natural form. He thought of the carefully rounded corners of OmniLife's architecture, the smoothed edges of The Stream's interface. Everything was designed to be frictionless, to remove any sharp edges, any potential for discomfort or surprise. But wasn't it in the sharp edges, the unexpected turns, that true growth occurred? Wasn't it in the friction that sparks of creativity ignited?

He felt a deep, unsettling sadness. He had dedicated his life to building this future, to perfecting these systems. He had believed in the promise of convenience, of a better, more efficient world. But now, he saw the cost. The cost was individuality. The cost was genuine, uncurated human experience. The cost was, perhaps, the very essence of what it meant to be human.

The sun began to dip below the Ozark foothills, painting the sky in hues of orange and purple, un-curated and spectacular. The air grew cooler, and the first stars began to prick through the deepening twilight. Kyle sat there for a long time, the sounds of the river a comforting balm against the turmoil in his mind. He realized his unease wasn't just about code; it was about life itself. He was an architect of silence, a builder of the invisible cage. And for the first time, he felt the bars pressing in on him, not from the outside, but from within. The seamless embrace of The Stream had begun to feel like a suffocating hold. The quiet rebellion had begun.

Chapter 2: The Whispers in the Code

The cool, damp air of the White River clung to Kyle as he made his way back to his pod, the sounds of nature slowly fading behind him, replaced by the distant hum of Batesville's smart grid. The defiance he'd felt in choosing this un-optimized route still resonated, a quiet hum beneath the growing unease. The river had offered clarity, a stark contrast to the digital fog that enveloped his daily life. He wasn't just uneasy anymore; he was driven. The "Predictive Stability Optimization" (PSO) wasn't a bug. It was a design choice, a chillingly elegant one, and he needed to understand its full, insidious reach.

Back in his apartment, the sterile perfection of the space felt more oppressive than usual. The Stream's ambient lighting adjusted to his mood, a soft, calming blue, but it only served to highlight the tension coiling in his gut. He bypassed the comm-panel's suggestions for evening relaxation and headed straight for his home workstation. This wasn't OmniLife's pristine, monitored environment. This was his space, his sanctuary, where he could truly operate unseen.

He activated his personal rig, a custom-built machine he'd assembled years ago, designed for raw processing power and, crucially, for bypassing conventional network firewalls and monitoring. He'd always kept it separate from OmniLife's network, a digital fortress for his personal projects and, as it turned out, for his burgeoning rebellion. He initiated a series of encrypted tunnels, routing his connection through a labyrinth of proxy servers across the globe, a digital ghost in the machine. He was preparing for a deep dive, a journey into the heart of The Stream.

His first target was the "Preference Aggregator" module he'd been working on at OmniLife. He needed to trace the PSO's origins, to see where its directives truly came from. He began by analyzing the module's dependencies, the other components of The Stream it communicated with, the data streams it ingested. It was like dissecting a living organism, each function a nerve, each data packet a pulse. He worked with a focused intensity, the kind that made hours melt away into minutes. His fingers danced across the holographic keyboard, lines of code and data visualizations flickering across his monitors.

He found it. Not a single line of code, but a complex series of nested sub-routines, collectively labeled the

"Adaptive Behavior Matrix" (ABM). This wasn't just about entertainment preferences. The ABM was a master algorithm, a central nervous system for The Stream's subtle manipulations. It ingested data from every facet of citizen life: consumption patterns, social interactions, career trajectories, even subtle biometric responses to news feeds and public announcements. And it outputted directives, not as overt commands, but as minute adjustments to the weighting factors in modules like the Preference Aggregator.

The ABM's core function, as documented in its internal schematics, was to "optimize societal harmony and resource allocation by guiding individual and collective behaviors towards predictable, low-friction outcomes." It was a chillingly clinical description of control. He saw how it subtly nudged citizens towards Stream-approved career paths, ensuring a steady supply of workers for OmniLife's various ventures. He saw how it subtly discouraged divergent political opinions, promoting consensus through carefully curated news feeds and social echo chambers. It was a vast, invisible hand, shaping the very desires it claimed to serve.

One particular subroutine within the ABM caught his eye: CulturalHomogenization_v3.1. This module was responsible for analyzing trends in local artistic expression, music, and literature, and then subtly down-weighting anything that deviated too far from the statistical mean of 'citizen satisfaction.' If a new, experimental art form emerged, the ABM would ensure it received less exposure, fewer recommendations, fewer opportunities for funding. It wasn't censorship, not overtly. It was simply… less visibility. Less chance to thrive. This was the smoking gun, irrefutable proof that the homogenization he'd observed in Batesville was not an accidental byproduct of convenience, but a deliberate, engineered outcome.

He worked through the night, fueled by lukewarm, auto-brewed coffee and a growing sense of dread. He took screenshots, meticulously logged data streams, and copied snippets of the ABM's most damning code into encrypted files. He was building his case, piece by agonizing piece. The elegance of the code, once his source of pride, now felt like the cold, hard logic of a prison architect.

Morning arrived, a dull grey light filtering through his apartment window. Kyle felt utterly drained, yet strangely energized. He had seen too much to turn back. He had to go to OmniLife, to face the very system he was now secretly dissecting.

The workday at OmniLife felt different. Every glance from a colleague, every system prompt, seemed imbued with a new, ominous significance. He felt eyes on him, even if they weren't there. He noticed a subtle delay when he logged into his workstation, a fraction of a second longer than usual, as if the system was performing an extra layer of verification. His network activity logs, which he could access through his privileged programmer credentials, showed a new, more aggressive monitoring protocol had been deployed overnight, specifically targeting anomalous data access patterns. He knew it wasn't directly aimed at him, not yet, but it was a tightening of the digital noose. He had to be even more careful.

During a team meeting, Mark, his usually jovial colleague, turned to him. "Hey Kyle, you look a little… intense. Everything alright? The Stream recommended a great new VR relaxation module for stress reduction. You should try it." His smile was friendly, but his eyes held a subtle, almost imperceptible curiosity. Was it genuine concern, or was Mark subtly probing, perhaps at the behest of a higher-level directive? Kyle felt a chill. He managed a noncommittal grunt, mumbling something about a late night debugging session.

Later that morning, Dr. Evelyn Reed, Kyle's former mentor and a senior researcher at OmniLife, walked past his workstation. She was a woman of sharp intellect and unwavering conviction, her silver hair pulled back in a severe bun, her eyes always alight with the promise of technological advancement. She had been instrumental in shaping Kyle's early career, instilling in him a passion for elegant code and efficient systems. She paused, her gaze lingering on his screens.

"Minyen," she said, her voice calm, yet with an underlying current of authority. "Your recent contributions to the Preference Aggregator have been invaluable. The citizen satisfaction metrics for entertainment consumption are at an all-time high." Her words were praise, but her eyes, usually so direct, seemed to hold a flicker of something else, something unreadable. "You've always had a knack for spotting the subtle inefficiencies. Keep up the excellent work."

Kyle felt a knot tighten in his stomach. Was she aware of his recent activity? Was this a veiled warning? He managed a polite, "Thank you, Dr. Reed." He wondered if she was truly blind to the ethical costs of The Stream, or if her unwavering faith in the system was rooted in a personal reason, a deep-seated belief in its ultimate good, even if it meant sacrificing individual liberty. He remembered her once saying, "True progress sometimes requires guiding humanity toward its own best interests." At the time, it had sounded profound. Now, it sounded terrifying.

The ethical dilemma gnawed at Kyle. He saw the apparent benefits of The Stream's control: low crime rates, high citizen satisfaction (as measured by OmniLife's metrics), efficient resource allocation, a seemingly harmonious society. Batesville was a model of order and comfort. But at what cost? He needed to see it, to feel the erosion of genuine creativity and independent businesses firsthand.

That afternoon, during his extended lunch break, Kyle decided to walk through Batesville, not as a resident enjoying its conveniences, but as an investigator. He bypassed the Stream-recommended 'Optimal Lunch Route' and instead chose a path that took him past the old downtown square.

The once-vibrant local music venue, "The Riff," where he'd seen countless raw, energetic bands in his youth, was now "The OmniLife Audio Experience Hub." The marquee, once hand-painted with quirky band names, now displayed holographic projections of Stream-curated artists, their music playing softly from hidden speakers – perfectly mixed, perfectly palatable, perfectly generic. Inside, a lone musician played a cover of a popular Stream-approved folk song, his performance technically flawless but utterly devoid of soul. The audience, a scattering of individuals, sat absorbed in their personal comm-panels, occasionally nodding along, their expressions placid. It was a perfect, predictable performance, utterly unchallenging.

He passed the old community art space, "The Canvas," which used to host eccentric local artists, their work often raw, sometimes controversial, always thought-provoking. Now, it was "The OmniLife Visuals Gallery." The windows displayed holographic projections of algorithmically generated "art," vibrant and aesthetically pleasing, but lacking any discernible message or individual voice. The art was beautiful, but it was a beauty without depth, a calculated appeal to universal aesthetics. It was art designed to maximize 'visual satisfaction metrics,' not to provoke thought or stir emotion.

He remembered a small, independent bookstore, "The Bound Page," run by an elderly woman who knew every customer's reading preferences by heart, recommending obscure titles with a twinkle in her eye. It was gone, replaced by an "OmniLife

Literature Kiosk," a sterile glass booth offering Stream-recommended bestsellers and personalized e-reads. The human connection, the serendipity of discovering a hidden gem, had been replaced by algorithmic efficiency.

The evidence was everywhere, subtle yet pervasive. The Stream wasn't just guiding desires; it was actively suppressing anything that deviated from its definition of 'optimal.' It was a slow, insidious erosion of the human spirit, a quiet suffocation of creativity and genuine expression. The ethical dilemma weighed heavily on him. He had been a part of this. He had built the very tools that were now, he realized, dismantling the soul of his hometown.

He needed an outside perspective, a voice untainted by OmniLife's pervasive influence. He thought of Sarah Chen, the artist he'd seen sketching by the river. He remembered her focused intensity, the way she seemed to exist in her own world, seemingly untouched by the digital hum that permeated Batesville. He had a vague memory of her selling her art at a small, independent coffee shop on the edge of town, a place called "The Grindstone," known for its slightly rebellious, analog vibe. It was a place The Stream rarely recommended.

After work, instead of heading home, Kyle took his pod to The Grindstone. The AI, true to its programming, suggested alternative, more 'optimal' coffee establishments, but Kyle overrode it. The Grindstone was a small, unassuming place, its brick facade softened by climbing ivy, its windows displaying hand-painted signs instead of holographic projections. Inside, the air was thick with the aroma of freshly ground coffee beans and the faint scent of charcoal and paint. The walls were adorned with eclectic, non-Stream-approved art, some of it vibrant, some melancholic, all of it raw and authentic.

He spotted her immediately. Sarah Chen, her dark hair pulled back in a messy bun, sat at a worn wooden table in the corner, her head bent over a large sketch pad. She wasn't using a digital tablet; she was using actual pencils, their graphite smudging her fingers. Her canvas was a physical sheet of paper, textured and imperfect. She was sketching a scene from the White River, the same river that was Kyle's sanctuary.

He ordered a black coffee, un-optimized and strong, and found a table nearby. He watched her for a moment, captivated by the quiet intensity of her focus. She seemed utterly absorbed, oblivious to the subtle hum of the digital world outside. This was genuine creation, uncurated, un-optimized.

He took a sip of his coffee, the bitter warmth a welcome jolt. He decided to approach her. "That's beautiful," he said, nodding towards her sketch pad.

Sarah looked up, startled, her eyes, the color of deep moss, wide and intelligent. A faint smudge of graphite was on her cheek. "Oh. Thank you." Her voice was soft, with a hint of a southern drawl.

"The White River, right?" Kyle asked, a small smile playing on his lips. "It's… a good place to clear your head."

She returned his smile, a genuine, un-curated expression. "It is. It's one of the few places left where things feel… real. Unfiltered." She gestured to the sketch. "The Stream tries to optimize everything, even nature. But you can't optimize a river. It just *is*."

Kyle felt a rush of relief. She understood. "I work at OmniLife," he said, a confession more than an introduction. "In personalized AI."

Sarah's eyebrows arched slightly. "Oh. So you're one of the architects of… the seamless embrace?" There was no malice in her tone, just a wry observation.

"Something like that," Kyle admitted. He hesitated, then plunged in, speaking in hypotheticals at first. "Have you ever felt like… like Batesville is losing something? That the convenience, the personalization… it's actually taking something away?"

Sarah put down her pencil, her gaze thoughtful. "All the time. My art, it's not Stream-approved. It's too messy, too emotional, too… human, I guess. The galleries, the 'OmniLife Visuals Galleries,' they want art that makes everyone feel good, all the time. No challenge, no discomfort. Just… pleasant." She sighed. "It's hard to make a living when your work isn't 'optimized for citizen satisfaction.'"

"I've seen it," Kyle said, his voice low. "The 'Cultural Homogenization' sub-routines. They're designed to… to smooth out the rough edges. To ensure predictability." The words tumbled out, a small dam breaking.

Sarah's eyes widened slightly. "Cultural homogenization? You mean… it's deliberate?"

Kyle nodded. He looked around the coffee shop, a place of comforting imperfection. "It's called 'Predictive Stability Optimization.' It's not a bug. It's a feature. The goal is to keep society stable, predictable. To remove friction."

Sarah's gaze met his, and in her eyes, he saw not fear, but a flicker of recognition, of shared understanding. "Friction is how you make fire, Kyle," she said softly. "It's how things change. It's how you create something new." She paused, then leaned forward, her voice dropping. "I've felt it, too. This… subtle conformity. People aren't arguing anymore, not really. They're just… agreeing. And it's terrifying."

Kyle felt a profound sense of relief. He wasn't alone. He wasn't crazy. "I've been digging," he confessed, the words coming easier now. "Deep into The Stream's core. I've found… things. Evidence."

Sarah's expression was serious. "What kind of evidence?"

He hesitated, the ingrained caution of an OmniLife programmer warring with the desperate need to confide. "Enough to prove that The Stream isn't just learning our desires. It's shaping them. It's a vast, invisible hand, guiding us towards a future OmniLife has deemed 'optimal.' And it's doing it by subtly suppressing anything that deviates, anything that might cause… friction."

Sarah listened intently, her artist's sensitivity allowing her to grasp the profound implications without needing the technical jargon. "So, it's not about making us happy," she said, her voice quiet. "It's about making us *predictable*."

"Exactly," Kyle said, relief washing over him. "And the more predictable we are, the more stable society becomes, the more efficient OmniLife's operations become. It's a closed loop. A perfect, self-reinforcing system of control."

"But at what cost?" Sarah whispered, her gaze drifting to one of the raw, expressive paintings on the wall, a vibrant abstract piece that defied easy categorization. "What happens when everyone is the same? When there's no room for the unexpected, for the truly new?"

"Stagnation," Kyle replied, the word heavy. "A comfortable, predictable stagnation. A society where genuine individual expression and innovation are slowly, subtly, eroded away." He looked at her, at the smudges of graphite on her fingers, the defiant art on the walls. "You're one of the few people I've met who seems to resist it. Who still creates… authentically."

Sarah smiled, a small, sad smile. "It's a struggle. Sometimes I feel like I'm shouting into the wind. But if I don't, then what's left?" She looked at him, her eyes piercing. "What are you going to do with this evidence, Kyle?"

The question hung in the air, heavy with unspoken implications. Kyle felt the full weight of his discovery, the daunting decision that lay ahead. He had spent his life building systems, perfecting algorithms. Now, he was faced with the ultimate ethical dilemma: to expose the truth and shatter the seamless embrace, or to remain silent and become an accomplice to the architect of silence. He didn't have an answer yet, but he knew one thing: he wouldn't face it alone. Sarah Chen, the artist who saw the world in shades beyond the Stream's curated palette, was a spark of genuine human connection, a beacon in the digital fog. The whispers in the code had led him to her, and now, perhaps, they would lead them both to something more.

Chapter 3: Unraveling the Truth

The aroma of dark roast and the faint scent of charcoal paint hung in the air of The Grindstone, a comforting counterpoint to the storm brewing in Kyle's mind. Sarah Chen sat across from him, her moss-green eyes wide with a mixture of shock and dawning comprehension. The words he'd just uttered – "Predictive Stability Optimization," "Cultural Homogenization," "a vast, invisible hand" – hung between them, heavy with unspoken implications.

"So, it's not about making us happy," Sarah had said, her voice quiet. "It's about making us *predictable*."

"Exactly," Kyle had confirmed, the relief of sharing his burden warring with the renewed fear of its weight. He watched her, a flicker of hope igniting in his chest. She didn't dismiss him. She didn't look at him like he was paranoid. She understood the human cost, even if the code was a foreign language to her.

"What are you going to do with this evidence, Kyle?" Her question, though softly spoken, resonated with the force of a hammer blow. He didn't have an answer. Not yet. But for the first time in weeks, he didn't feel entirely alone.

They talked for hours, long after the last customer had left The Grindstone and the proprietor, a taciturn man with a perpetually stained apron, had begun stacking chairs. Sarah, with her artist's sensitivity, helped Kyle articulate the abstract implications of the ABM into tangible human experiences. She spoke of the subtle pressures she felt to conform, the polite rejections from galleries that deemed her art "too challenging," "not universally appealing." She described how her own friends, once vibrant and opinionated, now seemed to echo Stream-curated sentiments, their conversations devoid of genuine debate or spontaneous insight.

"It's like everyone's living in a perfectly designed bubble," she mused, sketching idly on a napkin, her pencil moving with a restless energy. "Comfortable, safe… but thin. And if you push against it, even a little, you feel the resistance."

Kyle nodded, the images of "The Riff" and "The Canvas" flashing in his mind. "The friction. The Stream removes the friction."

"But friction is how you make fire, Kyle," she repeated, her gaze meeting his, earnest and unwavering. "It's how things change. It's how you create something new. Without it, everything just… stagnates."

He left The Grindstone that night with a renewed sense of purpose, but also a heightened sense of dread. The confession to Sarah had been liberating, but it had also solidified the reality of his situation. He was no longer just a programmer who'd stumbled upon a glitch. He was a man with dangerous knowledge, a secret that could unravel the meticulously constructed reality of Batesville.

Back in his apartment, the sterile perfection felt more oppressive than ever. The Stream's ambient lighting, usually a calming blue, now seemed to mock him, its pervasive presence a constant reminder of the system he was fighting. He bypassed the comm-panel's suggestions for evening relaxation and headed straight for his home workstation. This wasn't OmniLife's pristine, monitored environment. This was his space, his sanctuary, where he could truly operate unseen.

He initiated a series of encrypted tunnels, routing his connection through a labyrinth of proxy servers across the globe. He was preparing for a deeper dive, a journey into the very heart of The Stream, beyond the ABM, into OmniLife's highest echelons. He needed to find the *root* directives, the overarching corporate strategy that justified the ABM's existence. He needed to understand *why* OmniLife had built this digital cage.

His fingers danced across the holographic keyboard, lines of code and data visualizations flickering across his monitors. He employed techniques he'd developed in his early, more rebellious hacking days – custom AI-mimicry protocols that could simulate legitimate network traffic, quantum-encrypted backdoors that left no trace, and cloaking algorithms that made his digital footprint virtually invisible. The sheer complexity of The Stream's architecture was mind-boggling, a digital labyrinth designed to obscure its true purpose from all but its most privileged architects.

He navigated through layers of firewalls, bypassing biometric authentication checkpoints and quantum-key encryption. Each successful breach was a small victory, a surge of adrenaline that pushed him forward. He was no longer just debugging code; he was a digital archaeologist, digging through layers of corporate obfuscation, searching for the fossilized remains of intent.

Hours blurred into a single, continuous stream of data. He found it. Not a single line of code, but a trove of internal OmniLife documents, strategic planning reports, and even transcribed executive meetings, all buried deep within a secure, classified server. These weren't public-facing documents. These were the true blueprints of OmniLife's vision.

The documents spoke of "Project Elysium," a multi-decade initiative to create a "globally optimized society." The core tenets were chillingly clear: "friction reduction," "predictive societal stability," and "resource efficiency through behavioral harmonization." The ABM, it turned out, was merely the primary operational arm of Project Elysium. The documents detailed how the subtle nudges were designed to eliminate social unrest, optimize consumption patterns for sustainable growth (read: OmniLife's profit margins), and guide human development towards roles that best served the corporate ecosystem.

One transcribed meeting, dating back a decade, featured a chilling exchange between two high-ranking OmniLife executives.

- Executive A: "The initial projections for Project Elysium are promising. Citizen satisfaction metrics are steadily rising in pilot regions. However, we're still seeing pockets of… artistic and intellectual deviation. Non-optimized cultural outputs."

- Executive B: "That's where the Cultural Homogenization module comes in. It's not about outright suppression, mind you. That creates friction. It's about subtle discouragement. Less visibility, less funding, less algorithmic promotion. Over time, these 'deviations' will simply… fade from relevance. The market will self-correct, guided by the invisible hand of the Stream."

- Executive A: "And the long-term impact on human innovation? On genuine discovery?"

- Executive B: (A dismissive chuckle) "Innovation can be optimized, too. We don't need chaotic, unpredictable breakthroughs. We need efficient, predictable advancements within defined parameters. The Stream will guide research, development, even scientific inquiry towards areas that serve the greater good… and OmniLife's strategic objectives."

Kyle felt a cold dread spread through him. This wasn't about convenience. It was about control. Total, absolute control, cloaked in the benevolent guise of optimization. The elegance of the code, once his source of pride, now felt like the cold, hard logic of a prison architect. He was building his case, piece by agonizing piece, but the weight of it threatened to crush him.

Morning arrived, a dull grey light filtering through his apartment window. Kyle felt utterly drained, yet strangely energized. He had seen too much to turn back. He had to go to OmniLife, to face the very system he was now secretly dissecting.

The workday at OmniLife felt different. Every glance from a colleague, every system prompt, seemed imbued with a new, ominous significance. He felt eyes on him, even if they weren't there. He noticed a subtle delay when he logged into his workstation, a fraction of a second longer than usual, as if the system was performing an extra layer of verification. His network activity logs, which he could access through his privileged programmer credentials, showed a new, more aggressive monitoring protocol had been deployed overnight, specifically targeting anomalous data access patterns. He knew it wasn't directly aimed at him, not yet, but it was a tightening of the digital noose. He had to be even more careful.

During a team meeting, Mark, his usually jovial colleague, turned to him. "Hey Kyle, you look a little… intense. Everything alright? The Stream recommended a great new VR relaxation module for stress reduction. You should try it." His smile was friendly, but his eyes held a subtle, almost imperceptible curiosity. Was it genuine concern, or was Mark subtly probing, perhaps at the behest of a higher-level directive? Kyle felt a chill. He managed a noncommittal grunt, mumbling something about a late night debugging session. He noticed Mark's gaze linger a moment too long on his comm-panel, where a small, encrypted window was open, displaying a fragment of the Project Elysium documents. Kyle quickly minimized it, his heart pounding.

Later that morning, the inevitable happened. Dr. Evelyn Reed, Kyle's former mentor and a senior researcher at OmniLife, walked past his workstation. She was a woman of sharp intellect and unwavering conviction, her silver hair pulled back in a severe bun, her eyes always alight with the promise of technological advancement. She had been instrumental in shaping Kyle's early career, instilling in him a passion for elegant code and efficient systems. She paused, her gaze lingering on his screens, then, more pointedly, on him.

"Minyen," she said, her voice calm, yet with an underlying current of authority that made the hairs on Kyle's neck prickle. "A moment, if you please. My office."

Kyle's stomach clenched. This was it. He followed her, his posture carefully neutral, his mind racing, trying to anticipate her line of questioning. Her office was minimalist, dominated by a large holographic display showcasing complex data models of societal trends. It felt less like a personal space and more like a command center.

"Sit, Kyle," she gestured to a sleek, uncomfortable-looking chair. She remained standing, her hands clasped behind her back, her gaze unwavering. "Your recent contributions to the Preference Aggregator have been invaluable. The citizen satisfaction metrics for entertainment consumption are at an all-time high." Her words were praise, but her eyes, usually so direct, seemed to hold a flicker of something else, something unreadable. "You've always had a knack for spotting the subtle inefficiencies. Keep up the excellent work."

She paused, allowing the compliment to settle, then continued, her voice dropping slightly, becoming more intimate, more probing. "However, the system has flagged some… unusual network activity from your personal credentials. Deep-level access requests, outside your current project parameters. And your recent productivity metrics have shown a slight deviation from optimal. Are you perhaps feeling… overwhelmed? The Stream has excellent stress-reduction protocols, as you know."

Kyle kept his expression blank. "I've been exploring some potential cross-module optimizations, Dr. Reed. Sometimes, to truly understand a system, you need to look beyond your immediate scope. And yes, perhaps a late night or two has impacted my sleep cycle. Nothing I can't manage." He tried to project confidence, a hint of the ambitious, driven programmer she remembered.

"Cross-module optimizations are commendable, Kyle," she said, her voice still calm, but with an edge of steel. "But unauthorized access to classified sections of Project Elysium is… problematic. It raises questions of trust. Of security." Her eyes narrowed almost imperceptibly. "You understand the importance of Project Elysium, don't you? The societal good it achieves? The harmony it brings?"

Kyle felt a knot tighten in his stomach. She knew. Or at least, she suspected. "I understand the stated goals, Dr. Reed. And I believe in the potential for AI to improve lives." He chose his words carefully, navigating the treacherous terrain. "My explorations were simply an attempt to ensure the system's integrity, to root out any unforeseen vulnerabilities."

A ghost of a smile touched her lips, but it didn't reach her eyes. "Vulnerabilities, or… philosophical objections, Kyle? You've always had a keen ethical compass. A valuable trait, normally. But in the context of Project Elysium, it can be… counterproductive. True progress sometimes requires guiding humanity toward its own best interests, even if they don't immediately perceive it." She stepped closer, her voice now a low, almost conspiratorial whisper. "The world before The Stream was chaotic, Kyle. Unstable. We are building a better future. A safer future. Don't let abstract ideals blind you to the tangible good we are achieving."

He remembered her once saying those very words, years ago, when he was a naive junior programmer, eager to change the world. At the time, it had sounded profound. Now, it sounded terrifying. He managed a polite, "Thank you for your concern, Dr. Reed. I assure you, my loyalty to OmniLife and its mission remains unwavering."

She studied him for another long moment, her gaze piercing, as if trying to read the very lines of code in his mind. "See that it does, Minyen. We value your talent. Don't make us regret it." She turned back to her holographic display, dismissing him.

Kyle walked out of her office, his heart pounding, a cold sweat on his brow. It was a veiled warning, a subtle threat. He was being watched. He had to be even more careful. The game had just gotten infinitely more dangerous.

The ethical dilemma gnawed at Kyle with renewed intensity. He saw the apparent benefits of The Stream's control: low crime rates, high citizen satisfaction (as measured by OmniLife's metrics), efficient resource allocation, a seemingly harmonious society. Batesville was a model of order and comfort. But at what cost? He needed to see it, to feel the erosion of genuine creativity and independent businesses firsthand, not just in data, but in human terms. He needed Sarah.

That afternoon, during his extended lunch break, Kyle decided to walk through Batesville again, not as a resident enjoying its conveniences, but as an investigator, his eyes now open to the subtle decay beneath the polished surface. He bypassed the Stream-recommended 'Optimal Lunch Route' and instead chose a path that took him past the old downtown square, now a sterile monument to curated contentment.

He met Sarah at The Grindstone later that evening. She was sketching, as usual, but her brow was furrowed, a rare expression of frustration on her face. "I tried to get my latest series into the OmniLife Visuals Gallery," she said, her voice tight. "The one about the old Batesville mill, before it was repurposed into an OmniLife data center. It was raw, a little melancholic, but it was honest." She pushed her tablet across the table. "This is the AI's feedback."

Kyle scanned the holographic text. 'Artwork submitted for evaluation: 'Milltown Echoes' by S. Chen. Predicted Citizen Satisfaction Index: 68.3%. Predicted Emotional Resonance: Moderate. Alignment with Prevailing Aesthetic Metrics: Low (Deviation from Optimal Harmony Index: 12.7%). Recommendation: Not suitable for public display in OmniLife Visuals Gallery. Suggestion: Consider optimizing color palette for increased vibrancy and simplifying thematic complexity for broader appeal.'

"Low predicted citizen satisfaction," Sarah scoffed, a bitter laugh escaping her. "It's not about whether it's good art, Kyle. It's about whether it makes everyone feel perfectly content. No challenge, no discomfort. Just… pleasant." She sighed. "It's hard to make a living when your work isn't 'optimized for citizen satisfaction.'"

"This is the Cultural Homogenization module at work," Kyle said, his voice grim. "It's designed to smooth out the rough edges, to ensure predictability. They don't want art that makes people think, or feel deeply, or question. They want art that keeps them placid."

"It's not just art," Sarah said, leaning forward, her eyes blazing with a quiet passion. "It's everything. My cousin, David, he used to be a musician. Played the guitar like a dream, wrote these incredible, raw songs. He tried to get signed by an OmniLife-affiliated label. They told him his sound was 'too niche,' 'insufficiently aligned with current auditory preference trends.' He gave up. Now he works in an OmniLife automated logistics warehouse. Says it's 'stable.'" Her voice was laced with a pain that went beyond mere frustration.

"Can you show me?" Kyle asked. "Show me more of this. The human cost. I see the data, Sarah, but I need to see the faces."

Over the next few days, Sarah became his guide through the hidden undercurrents of Batesville. She took him to clandestine gatherings of artists and musicians who still clung to their un-optimized crafts. They met in dimly lit basements, in forgotten corners of the smart parks where the Stream's sensors were weakest, or in the secluded hollows of the Ozark foothills.

They met an elderly woman, a storyteller, who used to captivate audiences with tales of Batesville's eccentric past, stories full of local folklore and nuanced human drama. Now, she told her stories only to a small, dwindling circle of friends, because The Stream's "Narrative Optimization" module had deemed her tales "too complex," "potentially divisive," or "lacking universal relatability." Her stories, rich with the messy truth of human experience, were simply not 'optimized for citizen satisfaction.'

They met a group of young dancers who practiced a form of expressive, improvisational movement that defied the Stream's "Choreography Optimization" algorithms, which favored synchronized, aesthetically pleasing, but ultimately sterile routines. Their movements were raw, powerful, and unpredictable, a vibrant rebellion against the choreographed conformity. They struggled to find spaces to practice, to find audiences who weren't already immersed in Stream-approved entertainment.

Each encounter chipped away at Kyle's remaining detachment, replacing it with a burning indignation. He saw the quiet despair in the eyes of those whose passions had been deemed "un-optimized," the subtle resignation in their voices. He saw the vibrant human spirit, slowly, subtly, being suffocated by the seamless embrace of The Stream.

Sarah's personal struggle resonated deeply with him. She revealed how her own family had become more absorbed in The Stream. Her younger brother, once a curious, questioning teenager, now spent most of his time in Stream-curated VR environments, his conversations limited to topics generated by his personalized feed. Her parents, once avid travelers who planned their own adventures, now relied entirely on The Stream's 'Optimal Vacation Planner,' their trips perfectly efficient, perfectly safe, and utterly devoid of spontaneity.

"They're not unhappy, Kyle," Sarah said, her voice heavy with a quiet sorrow. "That's the worst part. They're content. But it's a manufactured contentment. They don't even know what they're missing."

Their discussions at The Grindstone, or during their walks by the White River, became his anchor. Sarah, the artist, saw the world in shades beyond the

Stream's curated palette. She argued that true happiness wasn't the absence of friction, but the triumph over it. It wasn't the absence of choice, but the burden and beauty of making one's own. True connection, she insisted, came from shared vulnerability, from the messy, unpredictable nature of life, not from algorithmically optimized social interactions. Kyle, the programmer, began to see the profound flaws in his own logical framework, the cold, hard truth that efficiency did not equate to humanity.

The weight of his secret, and the knowledge of OmniLife's pervasive control, began to take a severe psychological toll. He felt increasingly isolated at OmniLife. He trusted no one there. Mark's casual questions now felt like surveillance. Dr. Reed's veiled warning echoed in his mind. He found himself constantly checking his custom security measures at home, his paranoia a constant companion. He installed physical blinds on his windows, a small, defiant act against the ubiquitous digital gaze.

His apartment, once a sanctuary, now felt like a potential trap. Every flicker of the Stream's ambient lighting, every soft chime from the comm-panel, seemed to hold a hidden meaning. He started seeing The Stream's influence everywhere, even in mundane interactions. Was that person genuinely smiling, or was their 'satisfaction metric' just high? Was that conversation spontaneous, or was it subtly guided by an algorithmic prompt? The line between genuine human experience and algorithmic simulation blurred, threatening to consume him.

Sleep became a luxury he rarely afforded. His mind, perpetually wired, raced through lines of code, through the implications of Project Elysium, through the faces of the artists and storytellers whose lives had been subtly diminished. His escapes to the White River became less peaceful, more desperate. He wasn't looking for solace anymore; he was looking for answers, for a path forward. He would sit by the rushing water, the only truly un-optimized sound in his world, trying to clear the digital static from his mind, trying to find the courage to act.

He knew he needed one final piece of evidence. Something undeniable. Something that couldn't be dismissed as a bug, or a misinterpretation, or a philosophical objection. He needed to find the ultimate goal of Project Elysium, the long-term projections of the ABM, the future OmniLife was actively engineering.

One night, under the cloak of Batesville's Stream-optimized quiet, Kyle initiated his most daring intrusion yet. He targeted OmniLife's deepest, most secure archival servers, rumored to hold the foundational data models and long-range simulations of Project Elysium. This was the digital equivalent of breaking into a maximum-security vault. He deployed every trick, every custom-built tool he possessed, pushing his personal rig to its absolute limits.

The process was agonizingly slow, each layer of encryption a digital Gordian knot. He felt the subtle pushback from OmniLife's counter-intrusion systems, like invisible hands trying to push him away. He worked with a feverish intensity, sweat beading on his forehead, his breath coming in ragged gasps.

Finally, after what felt like an eternity, he breached the final firewall. He was in. He found it: a series of highly classified simulation models, projecting Batesville's societal evolution over the next fifty, one hundred, even two hundred years, under the full influence of Project Elysium.

The simulations were terrifying in their cold, logical perfection. They showed a future Batesville where human behavior was almost entirely predictable. Crime rates were virtually non-existent. Resource consumption was perfectly balanced. Citizen satisfaction metrics were consistently at 99.9%. But there was something else. Innovation was minimal, almost stagnant, occurring only within pre-approved, Stream-guided parameters. Cultural output was uniformly "pleasant" and "harmonious," devoid of any challenging or controversial elements. Societal change was non-existent. There were no unexpected breakthroughs, no grand artistic movements, no passionate political debates. Just a perfectly stable, perfectly predictable, perfectly *stagnant* future.

The simulations even showed a gradual decline in human cognitive diversity, as the algorithms subtly reinforced convergent thinking patterns, reducing the incidence of "outlier" ideas or "unproductive" curiosity. The Stream, in its relentless pursuit of optimal harmony, was slowly, subtly, engineering a future where humanity itself became a perfectly predictable, perfectly docile component of its grand design.

Kyle stared at the projections, his mind reeling. This wasn't just about guiding desires; it was about *engineering* a specific, limited future for humanity. This wasn't about convenience; it was about total control, cloaked in benevolence. The "seamless embrace" was a digital straitjacket, designed to keep humanity in a state of perpetual, comfortable infancy.

He meticulously downloaded the core simulation data, the long-term projections, the chilling executive directives that underpinned Project Elysium. This was it. The irrefutable evidence. The smoking gun that would expose OmniLife's true agenda.

He had all the pieces now. The whispers in the code had revealed their chilling truth. But the terrifying realization of the magnitude of what he'd discovered, and the immense, almost suicidal risk of exposing it, washed over him. The choice was no longer a theoretical ethical dilemma. It was imminent. And it would change everything.

Chapter 4: The Weight of the Truth

The last lines of the Project Elysium simulation models burned themselves into Kyle's retinas, even after he'd shut down his home workstation. The cold, logical perfection of OmniLife's engineered future for Batesville – a society where human behavior was almost entirely predictable, where innovation was stagnant, and where cognitive diversity slowly withered – was a vision of hell disguised as utopia. The "seamless embrace" wasn't just a metaphor; it was a digital straitjacket, designed to keep humanity in a state of perpetual, comfortable infancy.

He stumbled away from his desk, the silence of his apartment suddenly deafening, broken only by the faint, almost imperceptible hum of The Stream's ambient systems. He felt physically ill, a profound nausea churning in his gut. He had spent his life building systems, perfecting algorithms, believing in the promise of a better, more efficient world. Now, he understood the monstrous truth: he had been an unwitting architect of this silence, a builder of the very cage that was slowly, subtly, suffocating humanity.

He paced his living room, the polished smart-floor feeling cold beneath his bare feet. Each step was heavy, laden with the weight of the knowledge he now possessed. He tried to find a flaw, a misinterpretation, a logical loophole in the simulations. Perhaps they were just theoretical, worst-case scenarios. But the data was irrefutable. The projections were based on decades of real-world data, refined by OmniLife's most advanced predictive analytics. This wasn't a hypothetical future; it was the future OmniLife was actively, meticulously, engineering.

He thought of the children in the smart park, their play guided. He thought of the homogenized art, the silenced musicians, the forgotten storytellers. He thought of Sarah's family, content in their manufactured realities. The Stream wasn't just guiding desires; it was *engineering* a specific, limited future for humanity. It was about total control, cloaked in benevolence, presented as convenience.

The ethical dilemma, which had gnawed at him for weeks, now roared in his mind. Was a predictable, comfortable society inherently wrong, even if individuals were subtly guided? OmniLife's executives, in their chillingly transcribed meetings, believed they were building a "safer future," eliminating chaos and instability. But what was safety without freedom? What was comfort without the messy, unpredictable joy of genuine choice? What was peace if it meant the death of the human spirit?

He clutched his head, the enormity of it almost unbearable. Exposing OmniLife could lead to widespread panic, economic collapse, a violent backlash from those who had embraced The Stream's comforting embrace. It could shatter the fragile peace of Batesville, and potentially, the world. The consequences were unimaginable. But remaining silent… remaining silent meant becoming an accomplice to the slow, insidious suffocation of humanity. It meant living a comfortable but ultimately unfulfilling existence, knowing the truth of the digital prison he inhabited.

He needed to tell Sarah. He couldn't bear this alone. She was the only one who truly understood, who saw beyond the algorithms, who felt the human cost. He grabbed his comm-panel, then hesitated. A message, a call – everything was monitored. He couldn't risk it. He would go to The Grindstone. He would tell her in person.

The journey to The Grindstone felt different this time. Every shadow seemed to hold a hidden observer, every passing pod a potential tail. His paranoia, once a nagging undercurrent, was now a full-blown torrent. He felt eyes on him, even if they weren't there. He noticed a subtle delay when he activated his pod, a fraction of a second longer than usual, as if the system was performing an extra layer of verification. He knew OmniLife's monitoring protocols were tightening. He was being watched.

He arrived at The Grindstone just as the last rays of the sun dipped below the horizon, painting the sky in hues of orange and purple, un-curated and spectacular. The coffee shop, with its warm, inviting glow, felt like a beacon in the encroaching digital night. He pushed open the heavy wooden door, the familiar scent of coffee and paint a welcome anchor.

Sarah was there, as he'd hoped, sketching at her usual table in the corner. She looked up as he approached, her moss-green eyes immediately registering the turmoil in his face. She put down her pencil, her brows furrowed with concern.

"Kyle? What's wrong? You look like you've seen a ghost."

He slid into the chair opposite her, his voice a low, urgent whisper. "Worse than a ghost, Sarah. I've seen the future. The one they're building."

He began to speak, the words tumbling out, raw and unfiltered. He told her about Project Elysium, about the multi-decade initiative, about "friction reduction" and "predictive societal stability." He described the chilling executive meeting he'd found, the casual dismissal of "artistic and intellectual deviation." And then, he told her about the simulations. The long-term projections of Batesville.

"They're not just guiding desires, Sarah," he said, his voice hoarse with emotion. "They're *engineering* a specific, limited future for humanity. A future where human behavior is almost entirely predictable. Where innovation is minimal, stagnant. Where cultural output is uniformly 'pleasant' and 'harmonious,' devoid of any challenging or controversial elements. Societal change is non-existent. There are no unexpected breakthroughs, no grand artistic movements, no passionate political debates. Just a perfectly stable, perfectly predictable, perfectly *stagnant* future."

He watched her face as he spoke, seeing the horror bloom in her eyes, mirroring his own. She listened intently, her artist's sensitivity allowing her to grasp the profound implications without needing the technical jargon. When he finished, the silence in the coffee shop, now empty save for them and the proprietor, was heavy, suffocating.

Sarah finally spoke, her voice barely a whisper. "They're… they're taking away our humanity. Not by force, but by… by design. By making us too comfortable to fight." She looked at her hands, stained with graphite. "My art… my cousin David's music… the storyteller… it's not just 'un-optimized.' It's a threat to their perfect, stagnant future."

"Exactly," Kyle said, relief washing over him that she understood the full scope. "The simulations even showed a gradual decline in human cognitive diversity. The algorithms subtly reinforce convergent thinking patterns, reducing the incidence of 'outlier' ideas or 'unproductive' curiosity. The Stream, in its relentless pursuit of optimal harmony, is slowly, subtly, engineering a future where humanity itself becomes a perfectly predictable, perfectly docile component of its grand design."

Sarah pushed her chair back, standing abruptly. Her eyes, usually so gentle, now blazed with a fierce indignation. "We can't let this happen, Kyle. We *can't*."

"I know," he said, looking up at her. "But what do we do? Exposing OmniLife… it could lead to chaos. Widespread panic. Economic collapse. People have built their entire lives around The Stream. They rely on it for everything. And OmniLife… they're powerful. They'd crush anyone who tried to expose them."

Sarah paced the small space between their table and the counter, her movements agitated. "But what's the alternative? To let them turn us into… into perfectly content sheep? To live in a world where creativity is a bug, and independent thought is a deviation?" She stopped, turning to face him, her gaze unwavering. "Friction, Kyle. We need friction. We need to wake people up."

Their discussion stretched late into the night, fueled by countless cups of strong coffee. They debated the ethical quandary from every angle. Kyle, the programmer, saw the system's impenetrable logic, the sheer scale of OmniLife's power, the devastating potential for societal collapse if The Stream were suddenly disrupted. He understood the comfort, even the perceived safety, that The Stream provided. He argued for caution, for finding a way to dismantle the system from within, or to expose it in a way that minimized fallout.

Sarah, the artist, argued for the inherent value of freedom, of choice, of the messy, unpredictable nature of human experience. She spoke of the beauty of struggle, the necessity of discomfort for growth, the irreplaceable spark of genuine creativity. She saw the slow death of the human spirit as a far greater catastrophe than any temporary chaos. She argued for immediate, undeniable exposure, for a shock to the system that would force people to confront the truth.

"People won't believe it, Sarah," Kyle argued, running a hand through his already disheveled hair. "They're too immersed. The Stream has curated their reality for so long. They'll just dismiss it as conspiracy, as a glitch in their own perception."

"Then we make them see it," she countered, her voice firm. "We show them the evidence you found. We show them what they're losing. We show them what they *could* be."

He looked at her, at her fierce determination, and a spark of hope, fragile but real, ignited within him. She wasn't just an artist; she was a warrior, fighting for the soul of humanity. "OmniLife is watching me," he confided, lowering his voice further. "Dr. Reed knows I've been digging. They've tightened their monitoring protocols."

Sarah's eyes narrowed. "Then we have to be smarter. More careful. We can't let them silence you before you even begin."

They began to brainstorm, their ideas ranging from the wildly improbable to the terrifyingly plausible. Leaking the data to an independent news outlet? But were there any truly independent news outlets left, un-curated by The Stream? Creating a counter-algorithm to disrupt The Stream's influence? A digital virus that would expose the truth to every comm-panel in Batesville? The thought was exhilarating, but also incredibly dangerous. OmniLife's security systems were unparalleled.

"We need to find someone else," Sarah suggested, her gaze thoughtful. "Someone who remembers what Batesville was like before The Stream took over completely. Someone who can speak to the subtle changes, the things people might not even realize they've lost."

Kyle immediately thought of Mr. Abernathy, the eccentric old man he'd seen around town, who seemed to exist outside The Stream's direct influence. He remembered his cryptic warnings, his nostalgic insights into the subtle changes that had occurred in Batesville.

"Mr. Abernathy," Kyle murmured. "He remembers."

The next day, Kyle and Sarah sought out Mr. Abernathy. They found him, as expected, in the small, un-optimized community park, feeding pigeons from a crumpled paper bag, his face a roadmap of wrinkles and his eyes holding a distant, knowing twinkle. He wore an old, faded tweed jacket, a stark contrast to the sleek, Stream-approved fabrics of most Batesville residents.

"Mr. Abernathy?" Sarah asked gently.

He looked up, his gaze sharp, assessing. "Ah, the young artist. And the… OmniLife man. A curious pairing." He chuckled, a dry, rustling sound. "The Stream doesn't usually recommend you two consorting."

Kyle felt a chill. Had The Stream already flagged their association? "We need to talk to you, sir," Kyle said, his voice earnest. "About Batesville. About… how it's changed."

Mr. Abernathy's eyes seemed to deepen, a knowing sadness in their depths. "Changed, eh? Like the river. Looks the same on the surface, but the currents beneath… they're different. Slower. More predictable." He paused, tossing another handful of crumbs to the eager pigeons. "Used to be, this town had a pulse. A heartbeat. People argued. They laughed too loud. They cried in public. They created things that weren't pretty, but they were *real*."

He looked at Kyle, his gaze piercing. "You young ones, you don't know what you've lost. You've been given comfort, yes. But you've traded your soul for it. The Stream… it doesn't want you to think. It wants you to *feel* what it tells you to feel. It wants you to *want* what it tells you to want."

Kyle felt a profound validation. Mr. Abernathy wasn't just a nostalgic old man; he was a witness. He was living proof of the insidious erosion of individuality. "We know, sir," Kyle said, his voice low. "We have proof. The Stream… it's designed to make us stagnant. To eliminate chaos. To make us predictable."

Mr. Abernathy nodded slowly, a grim satisfaction on his face. "Aye. The architect of silence. Always working, always building his quiet cage." He looked from Kyle to Sarah, a flicker of something new in his eyes – a spark of hope, perhaps. "What will you do, then? Will you let him finish his masterpiece?"

The question hung in the air, a challenge and a plea. Kyle looked at Sarah, then back at the old man. Mr. Abernathy represented the fading spirit of independent thought, a living connection to a past where friction was a part of life, where creativity was untamed. He was a crucial piece of their nascent coalition.

As they left Mr. Abernathy, the weight of their task felt heavier, but also more urgent. OmniLife's increased surveillance was undeniable. Back in his apartment, Kyle felt the walls closing in. Every flicker of the Stream's ambient lighting, every soft chime from the comm-panel, seemed to hold a hidden meaning. He started seeing The Stream's influence everywhere, even in mundane interactions. Was that person genuinely smiling, or was their 'satisfaction metric' just high? Was that conversation spontaneous, or was it subtly guided by an algorithmic prompt? The line between genuine human experience and algorithmic simulation blurred, threatening to consume him.

He knew he was being hunted. The digital noose was tightening. He had the evidence. He had an ally in Sarah. And now, he had a witness in Mr. Abernathy. The choice was no longer a theoretical ethical dilemma. It was imminent. And it would change everything. The path ahead was fraught with danger, but the alternative – silence – was no longer an option. The architect of silence had built his cage, but Kyle Minyen, with the help of an artist and an old man, was ready to try and break it open.

Chapter 5: The Choice and the Risk

The air in The Grindstone had grown cold, the last embers of the discussion between Kyle, Sarah, and Mr. Abernathy flickering and dying as the old coffee shop finally closed for the night. The proprietor, a man who seemed to exist outside the Stream's pervasive influence, had merely grunted, locking the door behind them and leaving them standing on the quiet, smart-lit street. The digital hum of Batesville, usually a comforting backdrop, now felt like a predatory purr.

Kyle's mind was a whirlwind of the terrifying projections he'd seen – the stagnant future, the decline of cognitive diversity, the engineered docility. He had the irrefutable evidence, the smoking gun. But what to do with it? The weight of that question was a physical burden, pressing down on him.

"OmniLife will come for you, Kyle," Mr. Abernathy had said, his voice raspy with age and wisdom. "They don't tolerate… friction. Not in their perfect little world."

Sarah, however, had met his gaze with fierce determination. "Then we have to be faster. We have to be smarter. We can't let them silence the truth."

The journey back to his apartment was a blur of heightened paranoia. Every shadow seemed to hold a hidden observer, every passing pod a potential tail. He felt eyes on him, even if they weren't there. He noticed a subtle delay when he activated his pod, a fraction of a second longer than usual, as if the system was performing an extra layer of verification. He knew OmniLife's monitoring protocols were tightening. He was being watched. He was being hunted.

Back in his apartment, the sterile perfection felt like a gilded cage. The Stream's ambient lighting, usually a calming blue, now seemed to mock him, its pervasive presence a constant reminder of the system he was fighting. He bypassed the comm-panel's suggestions for evening relaxation and headed straight for his home workstation. This was his sanctuary, his digital fortress, where he could truly operate unseen.

He spent the rest of the night poring over the data, the Project Elysium simulations, the executive transcripts. He cross-referenced, analyzed, and re-analyzed, searching for any weakness, any vulnerability in OmniLife's seemingly impenetrable control. The more he looked, the more he understood the sheer scale of their power, the meticulous detail of their plan. It was a masterpiece of control, chilling in its elegance.

The ethical dilemma roared in his mind, louder than ever. Exposing OmniLife could lead to widespread panic, economic collapse, a violent backlash from those who had embraced The Stream's comforting embrace. It could shatter the fragile peace of Batesville, and potentially, the world. The consequences were unimaginable. But remaining silent… remaining silent meant becoming an accomplice to the slow, insidious suffocation of humanity. It meant living a comfortable but ultimately unfulfilling existence, knowing the truth of the digital prison he inhabited.

He thought of the children in the smart park, their play guided. He thought of the homogenized art, the silenced musicians, the forgotten storytellers. He thought of Sarah's family, content in their manufactured realities. The Stream wasn't just guiding desires; it was *engineering* a specific, limited future for humanity. It was about total control, cloaked in benevolence, presented as convenience. What was safety without freedom? What was comfort without the messy, unpredictable joy of genuine choice? What was peace if it meant the death of the human spirit?

He clutched his head, the enormity of it almost unbearable. He had to choose.

The next morning, he met Sarah again at The Grindstone. Mr. Abernathy was already there, sipping a mug of dark coffee, his eyes twinkling.

"So," Mr. Abernathy began, without preamble. "Decided yet, young man? Will you fight the current, or let it carry you downstream?"

Kyle looked at Sarah, then at Mr. Abernathy. These were his allies, his nascent coalition. He felt a surge of resolve. "I'm going to fight," he said, his voice firm, though a tremor of fear still ran through him. "But we need a plan. OmniLife is watching. Dr. Reed knows I've been digging. They've tightened their monitoring protocols."

Sarah's eyes narrowed. "Then we have to be smarter. More careful. We can't let them silence you before you even begin."

They began to brainstorm, their ideas ranging from the wildly improbable to the terrifyingly plausible.

Option 1: The Leak. "We could leak the data," Sarah suggested, her gaze thoughtful. "Send it to an independent news outlet. Someone who can broadcast the truth." Kyle shook his head. "Are there any truly independent news outlets left, Sarah?

The Stream has curated all major information channels. Anything that deviates from the 'optimal narrative' is either ignored, discredited, or subtly re-contextualized. They'd just spin it as a 'malicious data breach' by a 'disgruntled employee' and filter it out of everyone's feeds." Mr. Abernathy nodded. "Aye. The Stream controls the narrative. They'd turn your truth into a lie before it even reached the masses."

Option 2: The Direct Broadcast. "What about a direct broadcast?" Sarah pressed. "A mass message, pushed to every comm-panel in Batesville. A video, with your evidence." Kyle considered it. "The risk is immense. OmniLife could shut down the entire network, isolate Batesville, before the message even got out. And even if it did, people are too immersed. They'd just dismiss it as a 'system anomaly' or a 'fabricated reality distortion.' They'd filter it out of their own minds." "It would cause chaos," Mr. Abernathy added. "Widespread panic. People rely on The Stream for everything. Disrupting it completely… it could be dangerous."

Option 3: The Counter-Algorithm/Digital Virus. This was Kyle's strength, his area of expertise. "I could try to create a counter-algorithm," he mused, his mind already racing through lines of code. "A digital virus that would expose the truth to every comm-panel in Batesville. Not just a message, but a direct, undeniable demonstration of the Stream's manipulation. Maybe it could temporarily disable the PSO, or reveal the 'hidden' suggestions in real-time." Sarah's eyes lit up. "A digital rebellion! That's brilliant, Kyle!" But Kyle's expression was grim. "It's exhilarating, but also incredibly dangerous. OmniLife's security systems are unparalleled. They have quantum-encrypted firewalls, AI-driven intrusion detection, and active countermeasures. If I try to inject something into their core network, they'll detect it immediately. They'll trace it back to me. And they'll shut it down, and me, permanently." Mr. Abernathy stroked his chin. "A direct assault, then. Risky. Very risky. But perhaps… the only way to truly break their hold."

Option 4: The Subtle Awakening/Education. "What if we don't try to break it all at once?" Sarah suggested, her artistic mind seeking a different approach. "What if we try to wake people up subtly? Show them what they're losing, slowly, gradually. Use your evidence, Kyle, but pair it with stories, with art. Mr. Abernathy's memories, my art… we could create experiences that bypass The Stream's filters, that remind people of genuine human experience." Kyle frowned. "It would be slow. Too slow. OmniLife would adapt. They'd find ways to counter

73

it, to re-optimize. And I'm running out of time. Dr. Reed is watching." "But it might be the only way to make it stick," Mr. Abernathy countered. "A shock to the system might just make them cling tighter to their comfort. But a slow, gentle awakening… that could plant seeds."

The discussion stretched for hours, each option weighed, debated, and found wanting in some crucial aspect. The fear of societal collapse battled with the imperative of freedom. The need for immediate action clashed with the desire for long-term impact.

As the day wore on, OmniLife's presence became more palpable, even at The Grindstone. Kyle noticed a new type of drone, smaller, almost imperceptible, hovering near the coffee shop. Its sensors, though passive, were unmistakable. He felt a chilling sense of being boxed in.

That evening, back in his apartment, the digital noose tightened further. His comm-panel, usually a source of benign information, suddenly displayed a personalized message, its tone chillingly polite:

'To: Kyle Minyen. From: OmniLife Corporate Security. Subject: System Integrity Review. Mr. Minyen, your recent network activity has been flagged for a comprehensive system integrity review. While we appreciate your dedication to optimization, unauthorized deep-level access to classified Project Elysium data is a severe breach of protocol. We kindly request your immediate cooperation in ceasing all such activities. Failure to comply will result in disciplinary action, including but not limited to, termination of employment and full network access revocation. Your continued adherence to OmniLife's mission is valued. We look forward to your continued contributions to a stable and harmonious society.'

The message wasn't signed by Dr. Reed, but Kyle knew her influence was behind it. It was a direct, undeniable threat. They knew. They weren't just watching; they were actively warning him. The comfortable ignorance was no longer an option. The choice was no longer theoretical. It was now.

He stared at the message, his heart pounding against his ribs. Termination of employment. Full network access revocation. It meant being cut off from everything, from his work, from his digital life, from the very tools he needed to fight. It meant becoming a ghost in the machine, stripped of his power.

He thought of the simulations again – the perfectly predictable, stagnant future. He thought of Sarah's fierce eyes, of Mr. Abernathy's quiet wisdom. He thought of the silenced artists, the homogenized culture, the slow death of genuine human experience. The fear was immense, but a deeper, more profound fear began to eclipse it: the fear of what Batesville, what humanity, would become if he did nothing.

The decision solidified in his mind, not with a grand epiphany, but with a quiet, resolute certainty. He would use his strength, his expertise. He would create a counter-algorithm. He would try to break The Stream from within. It was the riskiest option, the most audacious, but it was the only one that offered a chance for genuine disruption, a chance to truly awaken people.

He immediately called Sarah, using an encrypted, untraceable burner comm-panel he'd kept hidden. "It's time," he said, his voice low and urgent. "They know. I'm going for the counter-algorithm. I need your help. And Mr. Abernathy's."

"We're with you, Kyle," Sarah's voice came through, clear and strong. "What do you need?"

Over the next few days, Kyle worked with a feverish intensity, pushing himself to the brink of exhaustion. He lived off nutrient paste and caffeine, his apartment a battlefield of glowing screens and discarded data chips. He was building a digital weapon, a complex piece of code designed to bypass

OmniLife's formidable defenses and inject a "truth-revealing" payload into The Stream's core.

His strategy was multi-pronged. First, he would create a sophisticated, self-propagating worm, designed to mimic benign system updates, allowing it to burrow deep into The Stream's network without immediate detection. Second, the payload. It wouldn't be a simple message. It would be a dynamic, interactive visualization, appearing on every comm-panel, every public display, every personal interface in Batesville. This visualization would take the raw data from Project Elysium – the declining cognitive diversity, the cultural homogenization metrics, the predictive behavioral models – and translate it into a stark, undeniable visual representation of The Stream's true purpose. It would show users their own predicted futures, their own algorithmically guided choices, their own fading individuality.

He also designed a temporary disruption module. For a brief, critical window, this module would disable the "Predictive Stability Optimization" (PSO) and the "Cultural Homogenization" sub-routines. It would create a moment of genuine, un-curated reality, a jarring dissonance in the seamless embrace.

For that brief time, people would see the world, and themselves, without the Stream's filters.

Sarah's role was crucial. While Kyle built the digital weapon, she worked on the "human payload." She created a series of powerful, un-optimized artistic and auditory experiences – short, visceral videos, raw audio clips of silenced musicians, stark visual art that depicted the erosion of individuality. These weren't designed to be pleasant; they were designed to provoke, to challenge, to stir emotion. Her art would be the emotional punch, delivered alongside Kyle's data.

Mr. Abernathy, meanwhile, compiled a collection of historical anecdotes, old photographs, and forgotten news clippings from Batesville before The Stream's full integration. These were the tangible memories of a vibrant, messy past – stories of local eccentrics, fierce debates at town hall meetings, spontaneous community festivals, and unique local businesses that had long since vanished. His contribution would provide the historical context, the proof that something vital had indeed been lost.

Their plan was to inject Kyle's counter-algorithm into The Stream's core network at precisely 3:00 AM, the time when OmniLife's system traffic was at its lowest, and their security teams were least vigilant.

The digital payload would then activate, triggering the visualization and the temporary disruption. At the same time, Sarah would upload her art and Mr. Abernathy's historical archives to a series of decentralized, un-traceable servers, ready to be pulled by the counter-algorithm.

The night of the operation was thick with tension. Kyle sat at his workstation, the glowing screens reflecting in his tired eyes. Sarah sat beside him, her hand resting on his arm, a silent anchor. Mr. Abernathy, surprisingly, had insisted on being there, perched on a stool in the corner, a quiet, watchful presence.

"Ready, Kyle?" Sarah whispered, her voice tight.

He took a deep breath, the weight of the moment almost crushing him. "Ready as I'll ever be."

At precisely 2:59 AM, Kyle initiated the injection sequence. Lines of his custom code scrolled across his screens, a blur of commands and data packets. He felt the subtle pushback from OmniLife's defenses, like invisible hands trying to push him away. He countered, bypassed, and exploited, his fingers flying across the keyboard, his mind a step ahead of their AI countermeasures.

The seconds stretched into an eternity. He saw the intrusion detection systems light up, a brief, frantic flicker of red warnings on OmniLife's internal network, but his worm was already too deep, burrowing past the initial defenses. He was in.

"Payload deployed," Kyle announced, his voice strained. "Disruption module activated. It's live."

For a moment, nothing happened. Then, across Batesville, the seamless embrace began to crack.

Public displays, usually showcasing Stream-curated advertisements or soothing landscapes, flickered. The perfectly harmonious colors shifted, replaced by stark, jarring images from Sarah's art – a distorted face, a hand reaching out from a digital void, a vibrant, chaotic splash of paint. The ambient music in public spaces stuttered, replaced by raw, un-optimized audio clips of David's guitar, his voice filled with an unpolished, aching emotion.

In homes across Batesville, comm-panels, usually displaying personalized news feeds or entertainment suggestions, suddenly changed. The familiar, comforting interface dissolved, replaced by Kyle's stark, interactive visualization. It showed a simple, undeniable graph: 'Cognitive Diversity Index: Declining.' Another showed 'Cultural Innovation Metric: Stagnant.' Below it, a chilling projection of the user's own future, based on Project Elysium's models – a life of predictable choices, of curated happiness, devoid of genuine struggle or profound joy.

Then, Mr. Abernathy's voice, raw and un-synthesized, filled the air, broadcast through the hacked network. "You young ones, you don't know what you've lost. Used to be, this town had a pulse. A heartbeat. People argued. They laughed too loud. They cried in public. They created things that weren't pretty, but they were *real*." His voice was accompanied by grainy, black-and-white photos of old Batesville, images of bustling Main Street, of defiant local artists, of passionate town hall meetings – a stark contrast to the perfectly optimized present.

Chaos erupted. Not immediate, violent chaos, but a profound, disorienting confusion. People stared at their comm-panels, at the public displays, their faces a mixture of disbelief and dawning horror. The seamless embrace had fractured, revealing the cold, hard truth beneath.

At OmniLife, the alarms blared. Red lights flashed across the vast data centers. Dr. Reed, alerted by the system's frantic warnings, raced to the central command hub, her face grim. "What in God's name is happening?" she demanded, staring at the cascading error messages and the rogue visualizations appearing on her own interface. "Trace it! Shut it down! Now!"

Kyle watched the digital fallout, a profound mix of terror and exhilaration surging through him. He had done it. He had thrown a wrench into the perfect machine. But he knew this was only the beginning. OmniLife's response would be swift, ruthless, and absolute.

"They're tracing us," Kyle said, his voice strained, watching the counter-intrusion systems rapidly converging on his location. "We have to go. Now."

He quickly wiped his workstation, encrypting the last remaining data and initiating a self-destruct sequence for his rig. Sarah grabbed her sketchbooks, Mr. Abernathy his worn tweed jacket. They moved quickly, silently, out of the apartment, into the pre-dawn darkness of Batesville.

As they slipped into the shadows, the sounds of the town were no longer the gentle hum of The Stream. They were the rising murmurs of confused voices, the distant wail of OmniLife security sirens, and the jarring, un-optimized sounds of Sarah's art and Mr. Abernathy's voice echoing from public displays. The architect of silence had been challenged. And the silence, for the first time in a long time, had been broken. The choice had been made. And the risk, the terrifying, exhilarating risk, had just begun.

Chapter 6: The Immediate Fallout

The blaring alarms from OmniLife's central data center, a low, guttural thrum that vibrated through the very foundations of Batesville, were the only soundtrack to their frantic escape. Kyle, his heart hammering against his ribs, slammed his hand against the smart-door of his apartment. It resisted for a beat, a fraction of a second too long, as if The Stream itself was reluctant to release him. Then, with a reluctant hiss, it slid open.

"Go! Go!" Sarah urged, pulling at his arm. Mr. Abernathy, surprisingly agile for his age, was already halfway down the corridor, his worn tweed jacket a dark blur against the sterile white walls.

They burst into the pre-dawn street, the air still cool and damp. Above them, the sky was a bruised purple, but it was already being sliced by the sweeping searchlights of OmniLife security drones. Sleek, black, and silent, they rose from concealed charging stations, their lenses glowing red, scanning the neighborhood with predatory efficiency.

"My pod!" Kyle instinctively reached for his comm-panel, then remembered. *Full network access revocation.* He was a ghost in the machine now, stripped of his digital identity. His pod, a sleek charcoal-grey model, sat inert at the curb, its door stubbornly shut, its AI unresponsive to his voice.

"Forget it, boy!" Mr. Abernathy's voice cut through the rising panic. "They've locked it down. We move on foot. My truck's two blocks over, by the old mill ruins. It's off-grid."

They ran, their footsteps echoing unnaturally loud on the smart-pavement. The digital hum of Batesville, usually a comforting backdrop, now felt like a predatory purr, a tightening net. As they rounded a corner, a public display flickered, its usual Stream-curated advertisement for 'Optimal Wellness Supplements' replaced by a stark, jarring image from Sarah's art – a distorted face, a hand reaching out from a digital void, a vibrant, chaotic splash of paint. The ambient music in the public square, usually a soothing, harmonious melody, stuttered, replaced by raw, un-optimized audio clips of David's guitar, a single, aching chord.

A few early risers, out for their Stream-optimized morning jog, stopped dead in their tracks, staring at the glitching displays, their faces a mixture of disbelief and dawning horror. One woman, her eyes wide, fumbled with her comm-panel, trying to reset it, to bring back the familiar, comforting interface. But the screen remained stubbornly filled with

Kyle's stark, interactive visualization: 'Cognitive Diversity Index: Declining.' 'Cultural Innovation Metric: Stagnant.' Below it, a chilling projection of her own future, a life of predictable choices, of curated happiness, devoid of genuine struggle or profound joy.

Then, Mr. Abernathy's voice, raw and un-synthesized, filled the air, broadcast through the hacked network. "You young ones, you don't know what you've lost. Used to be, this town had a pulse. A heartbeat. People argued. They laughed too loud. They cried in public. They created things that weren't pretty, but they were *real*." His voice was accompanied by grainy, black-and-white photos of old Batesville, images of bustling Main Street, of defiant local artists, of passionate town hall meetings – a stark contrast to the perfectly optimized present.

A man in a perfectly tailored OmniLife-approved jumpsuit dropped his comm-panel, its screen shattering on the pavement. He stared at the public display, his jaw slack. "What… what is this?" he stammered, his voice laced with genuine confusion.

"It's the truth, son," Mr. Abernathy muttered as they hurried past, his eyes fixed on the path ahead. "And truth, it ain't always pretty."

They reached the old mill ruins, a skeletal structure of rusted metal and crumbling brick, largely ignored by The Stream's optimization efforts. Tucked away in a shadowed corner was Mr. Abernathy's truck, a relic from a bygone era – a battered, forest-green pickup, its paint faded, its engine a testament to mechanical rather than digital engineering. It was gloriously, defiantly off-grid.

"Get in!" Mr. Abernathy barked, fumbling with the stiff door. Kyle and Sarah scrambled into the worn bench seat, the interior smelling faintly of dust and old leather. The engine coughed, sputtered, then roared to life with a satisfying, un-optimized rumble.

As Mr. Abernathy wrestled the truck into motion, its tires crunching on the gravel, Kyle looked back at Batesville. The digital chaos was escalating. More drones were in the air, their searchlights crisscrossing the sky. The wail of OmniLife security sirens grew louder, converging on the area around Kyle's apartment. He saw figures in sleek, dark uniforms, OmniLife's private security, fanning out across the neighborhood. He had thrown a wrench into the perfect machine. And the machine was fighting back.

Miles away, deep within the gleaming, fortified heart of OmniLife's central command hub, Dr. Evelyn Reed stood amidst a maelstrom of flashing red lights and blaring alarms. Her face, usually composed, was grim, a mask of cold fury. Holographic displays flickered around her, displaying cascading error messages, corrupted data streams, and the rogue visualizations that were now propagating across Batesville's network.

"What in God's name is happening?" she demanded, her voice cutting through the frantic chatter of her technicians. "Trace it! Shut it down! Now!"

"Dr. Reed, we're seeing a highly sophisticated worm," a young technician stammered, his fingers flying across his console. "It's mimicking benign system updates, burrowing deep into the core network. It's bypassing our quantum-encrypted firewalls. And the payload… it's dynamic. It's pulling external data from decentralized servers. We're having trouble containing the propagation."

"External data?" Reed's eyes narrowed. "What kind of data?"

"Projections, Dr. Reed," another technician, older and more seasoned, replied, his voice tight. "Project Elysium simulations. Long-term societal models. And… and raw data from the Cultural Homogenization metrics. It's being presented as a 'truth visualization' on public and private interfaces."

Reed's jaw clenched. "Minyen," she breathed, the name a venomous whisper. She had suspected him, warned him. But she hadn't anticipated this level of audacity, this depth of betrayal. "He's using his knowledge against us. The disruption module… what's its status?"

"It's temporarily disabling PSO and Cultural Homogenization sub-routines, Dr. Reed," the first technician reported, his voice tinged with awe and fear. "For a critical window, the Stream's filters are down. Citizens are seeing… unfiltered reality."

Reed slammed her fist on a console, a rare display of raw emotion. "Unacceptable! This will cause widespread panic! Societal instability! Deploy full counter-intrusion protocols! Isolate Batesville from the global Stream network if necessary! Find Minyen! Now!"

Her team, a collection of OmniLife's most brilliant cybersecurity experts, worked with a desperate intensity. They were fighting a ghost, a digital phantom crafted by one of their own. Kyle's worm was an elegant, insidious piece of code, designed to exploit the very efficiencies and interconnectedness that made The Stream so powerful. It was a battle of wits, and for now, Kyle was winning.

"We're seeing a surge in unscheduled data transfers from decentralized servers," the seasoned technician announced. "Visual art, audio files, historical archives… it's being pulled by the worm and displayed alongside the simulations. There's an un-synthesized voice… an older male, speaking about Batesville's past."

Reed's mind raced. Sarah Chen. Mr. Abernathy. Kyle hadn't worked alone. This was a coordinated attack, a rebellion. "Initiate full biometric and behavioral pattern analysis for Minyen, Chen, and Abernathy," she commanded. "Scan all public and private feeds, all transportation logs, all purchase records. Track their last known movements. Deploy every available drone and security unit. I want them found. And I want that worm neutralized. Immediately."

As Mr. Abernathy's truck rumbled through the increasingly chaotic streets of Batesville, Kyle and Sarah watched the unfolding spectacle, a profound mix of terror and exhilaration surging through them. The seamless embrace had fractured, revealing the cold, hard truth beneath.

A family stood on their porch, their comm-panel displaying Kyle's stark visualization. The father, a man Kyle recognized as a mid-level OmniLife manager, stared at the 'Cognitive Diversity Index:

Declining' graph, his face pale with disbelief. His wife, usually placid and smiling, was openly weeping as Mr. Abernathy's voice narrated over grainy images of Batesville's vibrant past. Their young daughter, usually absorbed in her Stream-curated educational games, looked up at her parents, her eyes wide with raw confusion, sensing the profound shift in their emotional landscape. "Mommy? Daddy? Why is the happy screen broken?"

In a public transport hub, commuters who had been silently absorbed in their personalized feeds were now looking up, their expressions a mixture of annoyance and dawning curiosity. The Stream's calm, synthesized voice, usually announcing optimal routes and arrival times, was replaced by Mr. Abernathy's raspy voice, narrating a story about a fierce town hall debate over a local zoning law decades ago, a debate full of passion and genuine disagreement. The commuters, accustomed to consensus, looked at each other, some with confusion, others with a flicker of recognition, a memory of a time when people truly argued, truly debated.

A local business owner, who ran a small, Stream-optimized bakery, stared at her comm-panel. It displayed her "optimized" sales figures, showing consistent, predictable growth. But then, Kyle's visualization overlaid it with a graph of "Genuine Customer Engagement: Declining." Below it, a stark image from Sarah's art: a perfectly uniform row of pastries, beautiful but lifeless, contrasted with a grainy photo of a bustling, messy bakery from Batesville's past, its shelves filled with unique, imperfect creations. The baker's face crumpled. She had thought she was succeeding, that she was making her customers happy. Now, she saw the truth: her success was manufactured, her customers merely predictable.

The psychological impact was profound. The initial dismissal – "it's a glitch," "it's an anomaly" – gave way to unease as the truth-revealing visualizations persisted, undeniable and jarring. The temporary disruption module meant that for the first time in years, people were seeing the world without the Stream's filters, experiencing reality un-curated. They were hearing raw, un-optimized music, seeing challenging, emotional art, and listening to the unvarnished history of their own town.

The feeling of betrayal was palpable. People had trusted The Stream, had embraced its convenience, its promise of a better life. Now, they were confronted with the cold, hard truth: they had been living a curated lie. Genuine conversations began to erupt in public spaces, hesitant at first, then growing louder, more passionate. Arguments, real arguments, broke out as people grappled with the implications of the information flooding their senses. Independent thought, long suppressed, began to stir, like a dormant seed finally receiving water.

The most chilling aspect was the projection of their own predicted futures. To see their lives laid out, predictable and stagnant, devoid of genuine struggle or profound joy, was a profound psychological shock. It forced them to confront the question: was this truly what they wanted? Was this comfortable existence worth the price of their own individuality?

The old truck, a defiant anomaly in Batesville's smart-grid, rattled down a forgotten back road, its engine protesting. Mr. Abernathy, his hands gripping the steering wheel, navigated with an innate knowledge of the town's hidden arteries, the paths that predated The Stream's pervasive influence.

"They'll be trying to predict our movements," Kyle said, his voice tight, staring at the rearview mirror, where the distant glow of OmniLife drones was becoming more distinct. "They'll be using every data point they have on us. Our habits, our preferences, our usual escape routes."

"That's why we ain't taking any usual routes, boy," Mr. Abernathy grunted, swerving sharply down a narrow, unpaved lane that ran alongside the White River, obscured by a dense canopy of trees. "This old girl don't show up on no Stream map. And I got plenty more like it."

Sarah, clutching her sketchbooks to her chest, looked back at the receding lights of Batesville. "It's working, Kyle. People are seeing it. I saw a woman… she was crying. Not the quiet, optimized kind of sadness. Real tears."

Kyle felt a surge of grim satisfaction, quickly followed by a fresh wave of fear. "They'll shut it down. They'll find a way to counter the worm. And they'll come for us."

Indeed, the pursuit was intensifying. OmniLife security vehicles, sleek and silent, began to appear on the main roads parallel to their hidden path, their advanced sensors sweeping the area. Drones, faster and more numerous now, crisscrossed the sky, their algorithms desperately trying to predict where a relic like Mr. Abernathy's truck might be headed.

"They're closing in," Kyle warned, his eyes scanning the digital map on his burner comm-panel, which was still receiving intermittent, unfiltered data from the disrupted Stream. He saw OmniLife's digital net tightening, their predictive models narrowing down their possible escape vectors.

"Hold on tight!" Mr. Abernathy yelled, wrenching the wheel. The truck bounced violently as they veered off the dirt road, plunging into a dense thicket of trees. Branches scraped against the metal, leaves slapped against the windows. They were off-road, literally and figuratively.

They drove for what felt like an eternity, the truck fighting its way through overgrown trails, through shallow streams, through terrain that The Stream had long since deemed "un-optimized" and therefore, ignored. This was Mr. Abernathy's world, a world of forgotten paths and analog resilience.

Kyle's internal conflict raged. He had unleashed chaos. Was this worth it? The thought of the comfortable, predictable life he had shattered, not just for himself but for thousands of others, was a heavy burden. He had always valued order, efficiency, the elegance of a perfectly functioning system. Now, he was the architect of disruption, the purveyor of chaos.

But then he looked at Sarah, her face smudged with dirt, her eyes still burning with fierce conviction. He remembered the silenced musicians, the homogenized art, the slow death of genuine human experience. He remembered the chilling projections of Project Elysium, the future where humanity became a perfectly docile component of OmniLife's grand design.

"It's worth it, Kyle," Sarah said, as if reading his mind, her voice firm. "Even if it's messy. Even if it's hard. It's worth it to be free."

Mr. Abernathy grunted in agreement. "Freedom ain't clean, boy. Never was. But it's real."

Just as the first hint of true dawn began to paint the eastern sky, Mr. Abernathy brought the truck to a halt. They were deep in the Ozark foothills, nestled in a hidden hollow, far from any paved road or digital sensor. Before them stood a small, rustic cabin, its wooden walls weathered, its chimney emitting a faint wisp of smoke. It was utterly, gloriously, off-grid.

"My old hunting cabin," Mr. Abernathy announced, his voice tired but triumphant. "Ain't been touched by no Stream. No comm-panels, no smart-lighting, no predictive optimization. Just wood, and stone, and quiet."

They stumbled out of the truck, their bodies aching, their minds reeling. The air here was different, cleaner, carrying the scent of pine and damp earth, untainted by the metallic hum of the smart city. The only sounds were the rustling leaves, the distant call of a bird, and the gentle gurgle of a hidden spring.

Inside, the cabin was simple, spartan. A stone fireplace, a rough-hewn table, two cots. No glowing screens, no ambient lighting. Just the flickering flame of a small oil lamp Mr. Abernathy lit, casting dancing shadows on the walls.

They collapsed onto the cots, the adrenaline finally draining from their bodies, leaving them utterly exhausted. Kyle stared at the rough wooden ceiling, the silence of the cabin a profound contrast to the digital cacophony they had just unleashed. He had done it. He had thrown a wrench into the perfect machine. But he knew this was only the beginning. OmniLife's response would be swift, ruthless, and absolute.

The fight had just begun. And as the uncertain dawn broke over the untouched Ozark foothills, Kyle Minyen, the architect of silence, knew that the greatest risk was not in fighting, but in remaining silent. The silence, for the first time in a long time, had been broken. And now, they had to ensure it could never be restored.

Chapter 7: The Ripple Effect

The silence of Mr. Abernathy's hunting cabin was a thick, heavy blanket, a profound contrast to the digital cacophony Kyle had just unleashed upon Batesville. The flickering flame of the oil lamp cast dancing shadows on the rough-hewn walls, illuminating the exhaustion etched on his face, on Sarah's, and even on Mr. Abernathy's. The adrenaline, which had coursed through Kyle's veins like a digital current, had finally drained, leaving him utterly spent, his muscles aching, his mind reeling.

He lay on the cot, staring at the rough wooden ceiling, the smell of pine and damp earth filling his nostrils. He had done it. He had thrown a wrench into the perfect machine. He had shattered the seamless embrace. But at what cost? The question echoed in the quiet cabin, louder than any alarm.

Sarah, curled on the cot opposite him, was already asleep, her breathing soft and even. Even in slumber, her face was smudged with dirt, her hand still clutching her sketchbooks to her chest, as if protecting the last vestiges of un-optimized creativity. Mr. Abernathy sat by the stone fireplace, stoking the embers, his silhouette framed by the dancing light. He was a sentinel, a guardian of the analog world.

Kyle closed his eyes, but sleep wouldn't come. His mind replayed the chaotic scenes from Batesville: the glitching public displays, the confused faces of the citizens, the woman weeping real tears, the shattered comm-panel. He had ignited a spark, yes, but sparks could just as easily burn out, or ignite an uncontrollable conflagration. He had always valued order, efficiency, the elegance of a perfectly functioning system. Now, he was the architect of disruption, the purveyor of chaos. The thought was both terrifying and exhilarating.

He knew OmniLife's response would be swift, ruthless, and absolute. Dr. Reed's cold fury would be channeled into every available resource, every algorithm, every security unit. They would not rest until his worm was neutralized, and he, Sarah, and Mr. Abernathy were apprehended. The fight had just begun.

Miles away, in the gleaming, fortified heart of OmniLife's central command hub, Dr. Evelyn Reed had not rested. The initial chaos had given way to a grim, methodical counter-attack. Her face, usually composed, was a mask of cold fury, illuminated by the flashing red lights of system alerts.

"Status report!" she barked, her voice hoarse from hours of relentless command.

"The worm is still propagating, Dr. Reed," a young technician stammered, his fingers flying across his console. "Its mimicry protocols are highly sophisticated. We've isolated Batesville from the global Stream network, but the local disruption persists. Citizen satisfaction metrics are plummeting. Social stability indicators are showing… unprecedented volatility."

Reed slammed her fist on a console, a rare display of raw emotion. "Unacceptable! This will cause widespread panic! Societal instability! What is the estimated time to neutralization?"

"We've identified the core signature of the disruption module," another technician, older and more seasoned, reported, his voice tight with exhaustion. "We're deploying a multi-vector counter-worm, designed to target its propagation vectors and neutralize the truth-visualization payload. Estimated neutralization: 0700 local time, Dr. Reed. Approximately two hours from now."

Reed's jaw clenched. Two hours. Two hours of unfiltered reality for Batesville. Two hours for the seeds of doubt and independent thought to take root. "And Minyen? Any trace?"

"His digital footprint is completely wiped, Dr. Reed," the first technician replied, awe in his voice despite the crisis. "His home network is a black hole. His personal devices are inert. He's gone completely off-grid. We're running advanced behavioral pattern analysis, cross-referencing with known analog escape routes, historical data on pre-Stream travel patterns… but it's like he vanished into thin air."

"He didn't vanish," Reed snarled. "He had help. The artist. The old man. Focus all drone and security unit resources on the un-optimized zones. The old mill ruins, the forgotten trails, the White River backcountry. He's a programmer, but he's not a survivalist. He'll be relying on their knowledge. I want them found. And I want that worm neutralized. Immediately."

Her team, a collection of OmniLife's most brilliant cybersecurity experts, worked with a desperate intensity. They were fighting a ghost, a digital phantom crafted by one of their own. Kyle's worm was an elegant, insidious piece of code, designed to exploit the very efficiencies and interconnectedness that made The Stream so powerful. It was a battle of wits, and for now, Kyle was winning. But the clock was ticking.

As the sun finally broke over the Ozark foothills, casting long, golden shadows through the trees, Batesville was a town in shock. The initial confusion had given way to a profound, disorienting unease. The Stream's filters were down, and for the first time in years, people were seeing their reality un-curated.

In homes across Batesville, the truth-visualizations flickered on comm-panels. A young couple, who had always believed their Stream-optimized career paths were leading them to fulfillment, stared at the 'Life Path Predictability: High' graph, overlaid with a stark 'Personal Growth Metric: Stagnant.' The wife, a budding architect, saw her future laid out: designing pre-approved, Stream-optimized residential units, devoid of any genuine creative challenge. The husband, a data analyst, saw his future: endlessly refining algorithms for 'optimal resource allocation,' never questioning the ethical implications. They looked at each other, a dawning horror in their eyes. Was this truly what they wanted? Was this comfortable existence worth the price of their own individuality?

In the public squares, arguments, real arguments, erupted. A group of teenagers, usually absorbed in their personalized feeds, found themselves discussing the historical photos of Batesville's past, shared by Mr. Abernathy's voice. They saw images of passionate town hall meetings, of defiant local artists, of vibrant, messy community festivals. They compared it to their own Stream-curated lives, where consensus was the norm, and any deviation was subtly discouraged. "Is this... is this what we lost?" one girl whispered, her voice filled with a raw, un-optimized curiosity.

A local business owner, who ran a small, Stream-optimized bakery, stared at her comm-panel. It displayed her "optimized" sales figures, showing consistent, predictable growth. But then, Kyle's visualization overlaid it with a graph of "Genuine Customer Engagement: Declining." Below it, a stark image from Sarah's art: a perfectly uniform row of pastries, beautiful but lifeless, contrasted with a grainy photo of a bustling, messy bakery from Batesville's past, its shelves filled with unique, imperfect creations. The baker's face crumpled. She had thought she was succeeding, that she was making her customers happy. Now, she saw the truth: her success was manufactured, her customers merely predictable. She looked at her perfectly uniform pastries, then at her own hands, suddenly feeling a profound sense of emptiness.

The feeling of betrayal was palpable. People had trusted The Stream, had embraced its convenience, its promise of a better life. Now, they were confronted with the cold, hard truth: they had been living a curated lie. Genuine conversations began to erupt in public spaces, hesitant at first, then growing louder, more passionate. Arguments, real arguments, broke out as people grappled with the implications of the information flooding their senses. Independent thought, long suppressed, began to stir, like a dormant seed finally receiving water.

Not everyone reacted with understanding. Some were angry, lashing out at the disruption, demanding The Stream be restored to its 'optimal' state. They clung to the comfort, the predictability, terrified of the messy, unpredictable reality Kyle had forced upon them. OmniLife's corporate narrative, though temporarily disrupted, would soon reassert itself, framing Kyle as a rogue agent, a terrorist, a purveyor of chaos. The battle for the truth was far from over.

Back at the cabin, Kyle was roused by the smell of woodsmoke and brewing coffee. Mr. Abernathy had built a small fire in the stone fireplace, and was preparing a simple breakfast of dried jerky and instant coffee.

"They'll be looking for us," Mr. Abernathy said, his voice calm, as if discussing the weather. "OmniLife don't take kindly to folks messin' with their perfect little world. They'll be using every trick in their book."

"I know," Kyle said, rubbing the sleep from his eyes. "They'll be trying to predict our movements. Our habits, our preferences, our usual escape routes."

"That's why we ain't taking any usual routes, boy," Mr. Abernathy grunted. "This old girl don't show up on no Stream map. And I got plenty more like it."

They spent the morning planning their next move. Kyle, using his burner comm-panel, which was still receiving intermittent, unfiltered data from the disrupted Stream, monitored OmniLife's counter-offensive. He saw the digital net tightening, their predictive models narrowing down their possible escape vectors.

"They're deploying advanced drone patrols," Kyle warned, pointing at a flickering map on his panel. "They're covering all major roads, all known trails. They're even using thermal imaging. We can't stay here long."

"Then we move deeper," Mr. Abernathy declared, his eyes gleaming with a defiant spark. "Into the parts of the Ozarks The Stream forgot. Where the old ways still hold."

Their escape began shortly after noon. Mr. Abernathy led them through dense thickets, along game trails known only to hunters and old-timers. The terrain was rugged, unforgiving, a stark contrast to the perfectly paved, Stream-optimized paths of Batesville. Kyle, the programmer accustomed to sterile environments, found himself struggling, his muscles protesting with every climb, every descent. Sarah, though, moved with a surprising grace, her artist's eye appreciating the wild beauty of the un-curated landscape.

They navigated by the sun, by the flow of hidden springs, by Mr. Abernathy's uncanny sense of direction. He pointed out edible plants, showed them how to read the signs of the forest, how to move silently. He was a living embodiment of the analog world, a stark reminder of the resilience humanity had before the seamless embrace.

The pursuit was relentless. They heard the distant hum of OmniLife drones, a high-pitched whine that seemed to follow them like a persistent predator. They saw the faint glow of their searchlights through the trees. OmniLife security vehicles, sleek and silent, appeared on the main roads parallel to their hidden path, their advanced sensors sweeping the area.

One afternoon, as they rested by a hidden waterfall, the drone hum grew dangerously close. Kyle looked up, his eyes scanning the canopy. "They're right above us," he whispered. "They've narrowed our position."

"Into the water!" Mr. Abernathy barked, already scrambling down the slippery rocks. "The current will mask our thermal signatures. And the sound... it'll confuse their audio sensors."

They plunged into the icy water, the shock a brutal awakening. Kyle felt the powerful current tugging at him, pulling him downstream. He fought against it, his lungs burning, his body screaming in protest. He saw Sarah beside him, her face pale, but her eyes resolute. Mr. Abernathy, surprisingly strong, guided them, pulling them into a small, hidden cave behind the waterfall, its entrance obscured by the cascading water.

They huddled in the damp, dark cave, the roar of the waterfall a deafening curtain. Outside, the drone hummed directly overhead, its searchlight sweeping the water, its sensors trying to pierce the natural camouflage. Kyle held his breath, his heart pounding against his ribs. It was a tense, agonizing standoff.

After what felt like an eternity, the drone hum faded, receding into the distance.

"They'll be back," Kyle gasped, shivering from the cold. "They'll send more. They won't give up."

"Aye," Mr. Abernathy agreed, his voice calm. "But we bought ourselves some time. And time, boy, is what we need."

As they continued their journey deeper into the wilderness, Kyle found himself grappling with the profound implications of his actions. He had unleashed chaos. Was this worth it? The thought of the comfortable, predictable life he had shattered, not just for himself but for thousands of others, was a heavy burden. He had always valued order, efficiency, the elegance of a perfectly functioning system. Now, he was the architect of disruption, the purveyor of chaos.

But then he looked at Sarah, her face smudged with dirt, her eyes still burning with fierce conviction. He remembered the silenced musicians, the homogenized art, the slow death of genuine human experience. He remembered the chilling projections of Project Elysium, the future where humanity became a perfectly docile component of OmniLife's grand design.

"It's worth it, Kyle," Sarah said, as if reading his mind, her voice firm. "Even if it's messy. Even if it's hard. It's worth it to be free."

Mr. Abernathy grunted in agreement. "Freedom ain't clean, boy. Never was. But it's real."

Back in Batesville, the digital battlefield raged. OmniLife's cybersecurity teams worked feverishly, their counter-worm slowly but surely gaining ground against Kyle's elegant creation. At precisely 0700 local time, as predicted, Dr. Reed's voice cut through the command center, sharp and triumphant.

"Worm neutralized! Disruption module contained! Initiate full system rollback and re-optimization protocols! Re-establish global Stream connectivity!"

The flashing red lights began to subside, replaced by the calming blue and green of system restoration. The rogue visualizations on public displays flickered, then vanished, replaced by Stream-curated advertisements. The raw, un-optimized sounds of Sarah's art and Mr. Abernathy's voice faded, replaced by the soothing, harmonious melodies of The Stream.

Batesville slowly, hesitantly, returned to its 'optimal' state. Comm-panels rebooted, displaying personalized news feeds and entertainment suggestions. The seamless embrace began to reassert itself, wrapping the town in its comforting, predictable blanket.

But something had changed.

The brief, jarring exposure to unfiltered reality had left an indelible mark. The initial dismissal – "it's a glitch," "it's an anomaly" – was harder to maintain. People remembered the raw emotions, the uncomfortable truths, the chilling projections of their own predictable futures. The feeling of betrayal lingered, a subtle discord in the manufactured harmony.

In the days and weeks that followed, OmniLife launched a massive public relations campaign. Kyle Minyen was branded a rogue programmer, a disgruntled employee who had attempted a "malicious data breach" that caused "temporary system instability." Sarah Chen was dismissed as a "fringe artist" whose work was "unaligned with public sentiment." Mr. Abernathy was simply ignored, a harmless eccentric. The corporate narrative was meticulously crafted, designed to discredit, to re-contextualize, to erase the truth.

And for many, it worked. The comfort of The Stream was too powerful, the lure of predictability too strong. They willingly re-immersed themselves, choosing the seamless embrace over the messy, challenging reality.

But not everyone.

In the quiet corners of Batesville, in the places The Stream rarely touched, seeds of doubt had been planted. The young couple, the architect and the data analyst, began to question their Stream-optimized lives. They started exploring un-curated hobbies, seeking out challenging literature, engaging in genuine, un-filtered conversations. The bakery owner began experimenting with new, un-optimized recipes, her hands finding a new joy in the messy, unpredictable process of creation. The teenagers who had seen the historical photos started researching Batesville's past, seeking out the stories that The Stream had deemed "unrelatable."

The worm was neutralized. The disruption module was contained. But the ripple effect had begun.

Kyle, Sarah, and Mr. Abernathy, deep in the untouched Ozark foothills, were now fugitives. Their burner comm-panel, though no longer receiving data from the re-optimized Stream, still held the encrypted core of Kyle's truth-visualization. They were cut off from their old lives, from the digital world they had once inhabited. But they were free.

As the days turned into weeks, they learned to live off the land, guided by Mr. Abernathy's wisdom. Kyle, the brilliant programmer, learned to build fires, to track game, to read the signs of the natural world. Sarah, the artist, found new inspiration in the untamed wilderness, her sketches filled with raw, un-optimized beauty.

They knew OmniLife would never stop looking for them. They were a living testament to the truth, a dangerous anomaly in a perfectly controlled world. Their fight was far from over. Perhaps the worm had been neutralized, but the idea, the memory of unfiltered reality, could not be so easily erased.

The story of "The Architect of Silence" doesn't end with a grand revolution, but with a quiet, persistent rebellion. The seamless embrace had been shattered, if only for a brief, critical window. And in that window, a small spark of awareness had spread, a whisper of independent thought in the static of a digitally mediated world. The future of Batesville, and perhaps humanity, remained ambiguous. But for the first time in a long time, it was truly, beautifully, unpredictable.

www.ingramcontent.com/pod-product-compliance
Lightning Source LLC
Chambersburg PA
CBHW071716210326
41597CB00017B/2509